SIGNAL PROCESSING WITH ALPHA-STABLE DISTRIBUTIONS AND APPLICATIONS

Adaptive and Learning Systems for Signal Processing, Communications, and Control

Editor: Simon Haykin

Werbos / THE ROOTS OF BACKPROPAGATION: From Ordered Derivatives to Neural Networks and Political Forecasting

Nikias and Shao / SIGNAL PROCESSING WITH ALPHA-STABLE DISTRIBUTIONS AND APPLICATIONS

Krstić, Kanellakopoulos, and Kokotović / NONLINEAR AND ADAPTIVE CONTROL DESIGN

Diamantaras and Kung / PRINCIPAL COMPONENT NEURAL NETWORKS: THEORY AND APPLICATIONS

SIGNAL PROCESSING WITH ALPHA-STABLE DISTRIBUTIONS AND APPLICATIONS

Chrysostomos L. Nikias
Min Shao
University of Southern California

A Wiley-Interscience Publication
JOHN WILEY & SONS, INC.
NEW YORK / CHICHESTER / BRISBANE / TORONTO / SINGAPORE

This text is printed on acid-free paper.

Copyright © 1995 by John Wiley & Sons, Inc.

All rights reserved. Published simultaneously in Canada.

Reproduction or translation of any part of this work beyond that permitted by section 107 or 108 of the 1976 United States Copyright Act without the permission of the copyright owner is unlawful. Requests for permission or further information should be addressed to the Permissions Department, John Wiley & Sons, Inc., 605 Third Avenue, New York, NY 10158-0012.

Library of Congress Cataloging in Publication Data:
Nikias, Chrysostomos L.
    Signal processing with alpha-stable distributions and applications / Chrysostomos L. Nikias, Min Shao.
        p. cm. — (Adaptive and learning systems for signal processing, communications, and control)
    "A Wiley-Interscience Publication."
    Includes index.
    ISBN 0-471-10647-X (alk. paper)
    1. Signal processing—Mathematics. 2. Distribution (Probability theory) I. Shao, Min. II. Series.
TK5102.9.N543   1995
621.382'2—dc20   95-1482

Printed in the United States of America

10 9 8 7 6 5 4 3 2

To Our Parents
and
In Memory of Stamatis Cambanis

Contents

Preface   xi

1 Introduction   1

   1.1 Statistical Methods in Signal Processing / 1
   1.2 Statistical Models / 1
       1.2.1 The Gaussian Model / 1
       1.2.2 The Stable Model / 2
   1.3 The Moment Theory / 3
       1.3.1 Second-Order Moment Theory / 4
       1.3.2 Higher-Order Moment Theory / 4
       1.3.3 Fractional Lower-Order Moment Theory / 6
   1.4 Applications of Stable Distributions and Stable Signal Processing / 7
       Problems / 11

2 The Stable Distribution   13

   2.1 Introduction / 13
   2.2 The Characteristic Function / 13
   2.3 The Probability Density Function / 14
   2.4 An Efficient Method for Computing $S\alpha S$ Density Functions / 19
   2.5 Properties of Univariate Stable Laws / 20
   2.6 Multivariate Stable Distributions / 24
   2.7 Exponential Correlation Functions / 27
   2.8 Generation of Stable Random Variates / 27
   2.9 Conclusion / 29
       Problems / 30

3 Symmetric Stable Random Processes   31

   3.1 Introduction / 31
   3.2 Fractional Lower-Order Moments / 32
   3.3 Negative-Order Moments / 34
   3.4 Linear Space of Stable Random Variables / 35

3.5 Basic Types of Stable Processes / 37
  3.5.1 Sub-Gaussian Processes / 37
  3.5.2 Linear Stable Processes / 38
  3.5.3 Harmonizable Stable Processes / 42
3.6 Conclusion / 43
  Problems / 44

4 Covariation and Conditional Expectation — 46

4.1 Introduction / 46
4.2 Covariations and Their Properties / 46
4.3 Conditional Expectation and Linear Regression / 50
4.4 Complex Symmetric Stable Variables / 53
4.5 Conclusion / 54
  Problems / 55

5 Parameter Estimates for Symmetric Stable Distributions — 57

5.1 Introduction / 57
5.2 Method of Maximum Likelihood / 57
5.3 Method of Sample Fractiles / 58
5.4 Method of Sample Characteristic Functions / 61
5.5 Tests for Infinite Variance / 62
5.6 Simulations / 63
5.7 Alternative Methods Based on Negative-Order Moments / 66
5.8 Conclusion / 70
  Problems / 70

6 Estimation of Covariations — 72

6.1 Introduction / 72
6.2 Fractional Lower-Order Moment Estimator / 72
6.3 Screened Ratio Estimator / 73
6.4 Least-Squares Estimators / 74
6.5 Sampling Results and Some Comparisons / 75
6.6 Conclusion / 77
  Problems / 77

7 Parametric Models of Stable Processes — 79

7.1 Introduction / 79
7.2 Parameter Estimation of AR Stable Processes / 79
  7.2.1 Generalized Yule-Walker Equation / 80

        7.2.2   Least-Squares Method / 81
        7.2.3   Least Absolute Deviation Estimates / 82
        7.2.4   Sampling Results and Performance Comparisons / 83
   7.3 Parameter Estimation of ARMA Stable Processes / 83
        7.3.1   Higher-Order Yule-Walker Equation / 84
        7.3.2   Sampling Results and Performance Comparisons / 89
   7.4 Overdetermined Yule-Walker Equation and Singular
        Value Decomposition / 90
   7.5 The Yule-Walker Equation with Moments of Arbitrary Order / 94
   7.6 Conclusion / 95
        Problems / 95

8 Linear Theory of Stable Processes   98

   8.1 Introduction / 98
   8.2 The Minimum Dispersion Criterion / 99
   8.3 A Suboptimal State-Space Prediction / 100
   8.4 MD Prediction of Linear Stable Processes / 101
   8.5 Adaptive Wiener-Type Filters for Stable Processes / 102
   8.6 Identification of LSI Systems / 106
   8.7 Conclusion / 107
        Problems / 107

9 Symmetric Stable Models for Impulsive Noise   109

   9.1 Introduction / 109
   9.2 Filtered-Impulse Mechanism of Noise Processes / 110
   9.3 Characteristic Function of the Instantaneous Noise Amplitude / 112
   9.4 First-Order Statistics of Envelope and Phase of Narrowband Noise / 115
        9.4.1   Joint Characteristic Function of the In-Phase and
                Quadrature Components / 115
        9.4.2   Bivariate Isotropic Stable Distributions / 117
        9.4.3   First-Order Statistics of Noise Envelope and Phase / 118
   9.5 Amplitude Probability Distribution / 119
   9.6 Comparisons with Existing Models and Experimental Data / 122
   9.7 Conclusion / 127
        Problems / 128

10 Signal Detection in Stable Noise   130

   10.1 Introduction / 130
   10.2 Locally Optimum and Suboptimum Detectors / 130

10.3 Performance of Locally Optimum and Suboptimum Detectors for Stable Noise / 134

10.4 Asymptotic Probability of Error / 139

    10.4.1 Optimum Receiver / 139

    10.4.2 Linear (Gaussian) Receiver / 140

    10.4.3 Limiter Plus Integrator / 141

    10.4.4 Cauchy Receiver / 143

    10.4.5 A Note on the Exact Computation of the Probability of Error / 144

    10.4.6 Performance Comparisons / 145

10.5 Conclusion / 151

    Problems / 152

11 Current and Future Trends in Signal Processing with Alpha-Stable Distributions   154

Appendix   157

Bibliography   159

Index   167

Preface

The Gaussian (normal) distribution plays a predominant role in signal processing. Its simple justification by the central limit theorem, attractive analytic properties, and ability to lead to simple solutions have established it as *the* most important statistical distribution in almost every aspect of signal processing as well as in many other areas of engineering and science. Therefore, it is not surprising that until recently, non-Gaussian signal processing has received little attention.

One of the main difficulties in non-Gaussian signal processing is computational complexity. Signal processing under the non-Gaussian assumption is usually nonlinear and does not have closed-form solutions. The recent explosion in the speed and power of computers and the tremendous cost reduction in VLSI hardware, however, have made possible the implementation of sophisticated signal processing algorithms that would have been computationally unthinkable just a decade ago. It is now not only feasible but also necessary in many important applications to consider more realistic non-Gaussian models and avoid the loss of accuracy and/or resolution resulting from the oversimplified Gaussian assumption.

In this book, we describe a particular type of non-Gaussian models, those specified by the alpha-stable distributions. The class of alpha-stable distributions is a direct generalization of the Gaussian distribution and shares many of its familiar properties such as the *stability property* and the *central limit theorem*. Since their discovery by Paul Lévy in 1925, a vast amount of knowledge has been accumulated about the properties of these probability distributions. They have been found to provide useful models for phenomena observed in such diverse fields as electrical engineering, economics, physics, hydrology, and biology. However, they still remain obscure to many researchers and engineers in signal processing. The purpose of this book is to provide an introduction to the stable distributions and processes and emphasize their applications in signal processing. We hope that this book will aid engineers and signal analysts in utilizing the stable laws to model and analyze non-Gaussian phenomena and stimulate further research in this area.

Our discussions will be focused mainly on the statistical properties, methods, and applications of symmetric alpha-stable distributions. The practical rather than theoretical aspects are emphasized. Consequently, this book should appeal to both researchers and practicing engineers in signal processing. The reader is assumed to have an introductory signal processing background. The extensive (although by no means exhaustive) bibliography given at the end of the book will be of value to those readers who wish to pursue certain topics in depth.

Chapter 1 presents an introduction to statistical methods in signal processing with an emphasis on both the Gaussian and alpha-stable models. The chapter extends the discussion to moment theory: second-order, higher-order, and fractional lower-order (order less than second) and their relationships to Gaussian and non-Gaussian models. Applications of stable distributions and stable signal processing methods are also described along with the historical development of this field in the past seventy years.

Chapter 2 provides an introduction to the stable distribution, its characterization and statistical properties. Efficient methods for computing alpha-stable density functions are described along with algorithms for generating stable random variables. A lot of the properties of stable laws discussed in this chapter serve as the basis for the signal processing methods and their applications presented in Chapters 5 through 10.

Chapter 3 introduces the concept of the fractional lower-order moments—both positive- and negative-order moments along with a description of three basic types of stable processes and their properties; namely, sub-Gaussian processes, linear processes based on MA, AR, and ARMA representations, and harmonizable processes.

The concept of covariation and statistical conditional expectation of symmetric alpha-stable random variables is the subject of Chapter 4. The definition and key properties of complex symmetric stable random variables are also presented in this chapter.

In Chapter 5, methods of estimating the parameters of a stable distribution (e.g., characteristic exponent alpha, dispersion, location parameter) from sample data are introduced. In particular, numerical methods based on maximum likelihood, sample fractiles, sample characteristic functions, and negative-order moments are described along with tests for infinite variance.

Chapter 6 introduces three different conventional methods for the estimation of covariations; namely, the fractional lower-order moment estimator, the screened ratio estimator, and the least-squares estimator.

In Chapter 7, we describe methods for the estimation of AR and ARMA model parameters based on covariations of the output observations. Results from several examples and performance comparisons among these methods are also presented.

The attempts that have been made to generalize the second-order linear theory for Gaussian processes to the lower-order linear theory for alpha-stable processes are described in Chapter 8. In particular, we present methods based on the minimum dispersion criterion for linear predictor, filtering, and system identification. The adaptive Wiener-type filters for stable processes and their signal processing applications is the last topic covered in this chapter.

Chapter 9 establishes that under appropriate assumptions, the impulsive noise processes obey the symmetric stable distribution. In the case of narrowband impulsive noise reception, we show that the joint distribution of the in-phase and quadrature components is isotropic stable. This chapter also shows that the symmetric stable model is much simpler and mathematically more appealing than Middleton's statistical-physical canonical models.

The problems of designing and implementing optimum and suboptimum signal detectors in the presence of additive alpha-stable noise is treated in Chapter 10. In particular, four different classes of alpha-stable receivers are described along with their performance assessment and applications.

Finally, Chapter 11 provides an overview of current application developments and future research trends of alpha-stable distributions. In particular, a brief description of ongoing work is given in detection, blind channel identification and deconvolution, robust beamforming, adaptive interference mitigation, and their application to communications, sonar, radar, as well as other engineering areas.

The authors are grateful to the many people who have contributed in making the completion of this book possible. Xinyu Ma, Dr. George A. Tsihrintzis, Jun Shen, Dr. Robert Pierce, Dr. Dae C. Shin, and Panagiotis Tsakalides carefully read the manuscript and provided important feedback. We would like to thank Dr. George A. Tsihrintzis for permission to include some of his work on signal detection in this book as well as for his detailed review of the entire manuscript. We are particularly grateful to Xinyu Ma for developing the homework problems, editing, and correcting the book. The first author would also like to acknowledge gratefully the research support of the U.S. government, in general, and in particular, the Office of Naval Research, which funded part of the research on which this book is partially based.

CHRYSOSTOMOS L. NIKIAS
Rancho Palos Verdes, CA

MIN SHAO
San Diego, CA

1

Introduction

1.1 STATISTICAL METHODS IN SIGNAL PROCESSING

One of the basic objectives of signal processing is to extract desired information from observed data (signals). Since in most cases, the signals are either nondeterministic or contaminated by random noise, mathematical statistics plays an important role in signal processing [Nikias and Petropulu, 1993; Scharf, 1991; Shiavi, 1991; Mendel, 1987]. The application of statistical theory to signal processing typically involves the use of a probabilistic model to describe the way in which the observed signals and noise are generated. The probabilistic model for the underlying signals and noise is usually a function of the desired information, which is often a set of model parameters. This set of parameters is then determined from the data by using certain optimality criteria; obviously, the accuracy of the information extracted from the data in this way depends heavily on the probability model and optimality principles used in the inference.

1.2 STATISTICAL MODELS

In statistical signal processing problems such as signal detection and parameter estimation, detailed statistical models for both signals and noise are necessary. Good statistical models should be both realistic and relatively easy to analyze.

1.2.1 The Gaussian Model

The signal processing literature has traditionally been dominated by the Gaussian model. In many instances the Gaussian assumption is reasonable and can be justified by the Central Limit Theorem. Also, it often leads to analytically tractable solutions for signal processing problems. For example, the additive white Gaussian noise assumption in communication theory greatly simplifies the design and analysis of receiver structures. Any non-Gaussian assumptions will usually introduce nonlinearities.

Unfortunately, many signals and noise sources encountered in practice are decidedly non-Gaussian [Wegman and Smith, 1984; Kassam, 1988; Wegman, Schwartz, and Thomas 1989]. For example, underwater acoustic signals, low-frequency atmospheric noise, and many types of man-made noise are found to be non-Gaussian [Bouvet and Schwartz, 1989; Mertz, 1961; Lerner, 1961; Shinde and Gupta, 1974; Machell, Penrod, and Ellis, 1989; Middleton, 1977]. Non-Gaussianity often results in significant performance degradation for systems optimized under the Gaussian assumption. A well-known example is the matched filter for coherent reception of a deterministic signal in Gaussian white noise. If the noise statistics deviate from the Gaussian model, serious degradation in performance occurs, such as increased false alarm rate or error probability [Seo, Cho, and Feher, 1989; Schwartz and Thomas, 1984; Izzo and Paura, 1981; Izzo, Panico, and Paura, 1982]. On the other hand, a modest degree of nonlinear signal processing based on the actual noise statistics can lead to a much better receiver than the matched filter [Middleton, 1979; Lerner, 1961; Pillai and Harisankar, 1987; McCain and McGillem, 1987]. Thus, there is a trade-off between model complexity and accuracy. When the loss of resolution or accuracy due to the ideal Gaussian assumption in a non-Gaussian environment cannot be tolerated, more realistic statistical models must be considered.

1.2.2 The Stable Model

A broad and increasingly important class of non-Gaussian phenomena encountered in practice can be characterized as impulsive. Signals and noise in this class tend to produce large-amplitude excursions from the average value more frequently than Gaussian signals. They are more likely to exhibit sharp spikes or occasional bursts of outlying observations than one would expect from normally distributed signals. As a result, their density functions decay in the tails less rapidly than the Gaussian density function. Underwater acoustic signals, low-frequency atmospheric noise, and many types of man-made noise have all been found to belong to this class [Bouvet and Schwartz, 1989; Mertz, 1961; Lerner, 1961; Shinde and Gupta, 1974; Machell, Penrod, and Ellis, 1989; Middleton, 1977]. It is for this type of signals and noise that the *stable* distributions provide a useful theoretical tool.

The stable law is a direct generalization of the Gaussian distribution and in fact includes the Gaussian as a limiting case.[1] The main difference between the stable and the Gaussian distributions is that the tails of the stable density are heavier than those of the Gaussian density. This characteristic of the stable distribution is one of the main reasons why the stable distribution is suitable for modeling signals and noise of impulsive nature. In addition, the stable distribution is very flexible as a modeling tool in that it has a parameter α ($0 < \alpha \leq 2$), called the *characteristic exponent*, that controls the heaviness of its tails. A small positive value of α indicates severe impulsiveness, while a value of α close to 2 indicates

---

[1]In this book, stable distributions are usually assumed to be non-Gaussian, although logically the Gaussian distribution belongs to the family of stable distributions.

a more Gaussian type of behavior. When $\alpha = 2$, the stable distribution is reduced to the Gaussian distribution.

Theoretical justifications for using the stable distribution as a basic statistical modeling tool come from the central limit theorem, just like in the case of the Gaussian distribution. Recall that the classical Central Limit Theorem states that a physical phenomenon is Gaussian if there are infinitely many independent and identically distributed (i.i.d.) contributing factors, each of finite variance. However, there is a more powerful theorem, called the generalized central limit theorem, for the sum of infinitely many i.i.d. random variables. This theorem states that if the sum of i.i.d. random variables with or without finite variance converges to a distribution by increasing the number of variables, the limit distribution must belong to the family of stable laws [Feller, 1966; Breiman, 1968]. Thus, non-Gaussian stable distributions arise as sums of random variables in the same way as the Gaussian distribution. If an observed signal or noise can be thought of as the sum or result of a large number of independent and identically distributed effects, then the Generalized Central Limit Theorem suggests that a stable model may be appropriate.

Another defining feature of the stable distribution is the so-called *stability property*, which says that the sum of two independent stable random variables with the same characteristic exponent is again stable with the same characteristic exponent. In addition, the stable distribution is the only distribution that has this property. Intuitively, the stability property is very desirable, especially in modeling random noise and uncertain errors.

1.3 THE MOMENT THEORY

Statistical moments of signals provide rich sources about the desired information. The whole spectrum of statistical moments runs from order 0 to order ∞ (see Fig. 1.1). Traditional signal processing methods utilize only the second-order moments. Recently developed higher-order statistical signal processing techniques extract useful information from third- and fourth-order statistics. In this book, we will show that signal processing methods based on stable models require fractional lower-order moments, i.e., moments of order less than 2.

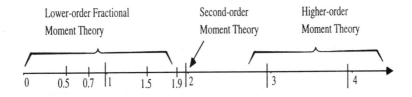

FIGURE 1.1 Statistical moments.

1.3.1 Second-Order Moment Theory

Suppose that $\{x(n)\}$ is a wide-sense stationary sequence with zero mean. The second-order moment (i.e., the *autocorrelation sequence*) defined by (where $\mathbf{E}[\cdot]$ is the statistical expectation operator)

$$R_x(k) = \mathbf{E}[x(n+k)x(n)] \tag{1.1}$$

together with its Fourier transform (i.e., the *power spectrum*)

$$S_x(\omega) = \sum_{k=-\infty}^{\infty} R_x(k)e^{-jn\omega} \tag{1.2}$$

are the foundation of statistical signal modeling and processing. Over the last fifty years, they have provided many important concepts and structures in signal processing, such as spectral representation, adaptive filtering, and linear prediction theory.

Based on the second-order moments, the least-squares criterion that minimizes the second-order moment of estimation errors has become the only natural choice for the optimality of signal processing methods. Although it is adequate under the Gaussian assumption and usually leads to analytically tractable results, the least-squares criterion is no longer appropriate in a non-Gaussian environment, largely due to its nonrobustness against a small number of big errors (outliers) in the data set [Bloomfield and Steiger, 1983; Gonin and Money, 1989]. When the least-squares criterion is used, little attention is paid to relatively minor errors in order to make very large errors as small as possible. In many situations in signal processing, however, it is more important to make as many errors small as possible, even if it is necessary to tolerate occasional large errors. When the error distribution is Gaussian, it does not matter which criterion is used because the most probable error is small. However, it can easily be demonstrated that least-squares estimates change dramatically when only a small proportion of extreme observations is present in the data [Tarantola, 1987]. In such cases, meaningful error criteria and robust procedures are needed [Denoel and Solvay, 1985; Claerbout and Muir, 1973; Taylor, Banks, and McCoy, 1979; Schroeder, 1991; Rice and White, 1964].

1.3.2 Higher-Order Moment Theory

For the past ten years non-Gaussian signal processing has received more and more attention in the literature. One main reason is recent advances in computer software and hardware. The tremendous growth in computational power, backed by cheap and fast VLSI hardware, makes it possible to implement very sophisticated signal processing algorithms. More realistic stochastic models can now be used to describe signals and noise. Another reason for the recent intense interest in non-Gaussian signal processing is the growing demand for products that serve the real world. The loss of resolution and accuracy due to the Gaussian assumption in a non-Gaussian environment is not always tolerable. Nowadays, when facing the trade-off between computational complexity of signal processing systems and realistic modeling of signals and noise, more often than not, signal processing engineers and scientists will choose the latter. These two main factors have led to a significant amount of research activities and applications in reexamining structures and inference methods in the context of filtering, estimation, detection, and signal extraction under non-Gaussian environments [Wegman and Smith, 1984; Wegman, Schwartz, and Thomas, 1989; Kassam, 1988].

For non-Gaussian signal processing problems, second-order descriptions do not provide enough information for understanding the non-Gaussianity involved. In this case, it has been suggested in the literature [Nikias and Petropulu, 1993; Nikias and Raghuveer, 1987] that one should look beyond the second-order statistics and extract phase information as well as information about deviations from Gaussianity from higher-order statistics, if they exist.

Although all moments of order greater than 2 are higher-order statistics, one usually focuses on the third- and fourth-order moments (cumulants) and their Fourier transforms (bispectrum and trispectrum respectively). Let $\{x(n)\}$ be a strictly stationary process with zero mean. Its third- and fourth-order cumulants are defined by

$$c_3(k_1, k_2) = \mathbf{E}[x(n)x(n+k_1)x(n+k_2)] \tag{1.3}$$

and

$$\begin{aligned} c_4(k_1, k_2, k_3) = &\; \mathbf{E}[x(n)x(n+k_1)x(n+k_2)x(n+k_3)] \\ &- \mathbf{E}[x(n)x(n+k_1)]\mathbf{E}[x(n+k_2)x(n+k_3)] \\ &- \mathbf{E}[x(n)x(n+k_2)]\mathbf{E}[x(n+k_1)x(n+k_3)] \\ &- \mathbf{E}[x(n)x(n+k_3)]\mathbf{E}[x(n+k_1)x(n+k_2)]. \end{aligned} \tag{1.4}$$

The bispectrum is defined by

$$C_3(\omega_1, \omega_2) = \sum_{k_1=-\infty}^{\infty} \sum_{k_2=-\infty}^{\infty} c_3(k_1, k_2) \exp[-j(k_1\omega_1 + k_2\omega_2)], \tag{1.5}$$

while the trispectrum is defined by

$$C_4(\omega_1, \omega_2, \omega_3) = \sum_{k_1=-\infty}^{\infty} \sum_{k_2=-\infty}^{\infty} \sum_{k_3=-\infty}^{\infty} c_4(k_1, k_2, k_3) \exp[-j(k_1\omega_1 + k_2\omega_2 + k_3\omega_3)]. \tag{1.6}$$

A well-known fact is that second-order descriptions or power spectrum suppress phase information, whereas the bispectrum and trispectrum preserve both magnitude and phase information of non-Gaussian signals [Lii and Rosenblatt, 1982]. Thus, in applications where accurate nonminimum phase information is desired, one has to utilize higher-order statistics. This is especially important in deconvolution problems that arise in geophysics, telecommunications, etc., in which the nonminimum phase of signals must be preserved. Other attractive features of higher-order spectra include their abilities to suppress Gaussian noise and detect and characterize nonlinearities in time series [Nikias and Petropulu, 1993; Nikias and Raghuveer, 1987].

Although higher-order spectral analysis is a powerful tool for non-Gaussian signal processing, it has its own limitations. First of all, higher-order spectral analysis is applicable only if the higher-order statistics exist. Although this may not seem to be a major problem, we will soon see that there are important applications where the underlying signals do not even have second-order moments. Secondly, there is still lack of robust procedures for estimating higher-order statistics from noisy observations. Finally, there is still no unifying theory for understanding higher-order spectra as a whole in the context of filtering, estimation, detection, and signal extraction. In particular, unlike the Gaussian case, signal processing methods based on higher-order statistics rarely have adequate optimality criteria. This, sometimes, makes it difficult to analyze and compare analytically signal processing algorithms based on higher-order statistics.

1.3.3 Fractional Lower-Order Moment Theory

From the signal processing point of view, the adoption of a stable model for signals or noise has important consequences. It is known that, for a non-Gaussian stable distribution with characteristic exponent α, only moments of order less than α are finite. In particular, the variance (i.e., the second-order moment) of a stable distribution with $\alpha < 2$ does not exist, making the use of variance as a measure of dispersion meaningless. Similarly, many standard signal processing tools (e.g., spectral analysis and all least-squares techniques) that are based on the assumption of finite variance will be considerably weakened and may, in fact, give misleading results.

The question on the role of the least-squares criterion and technique under the stable model assumption is not unexpected. Recall that the stable distribution is best used to model signals and noise that exhibit impulsive nature. This type of signal tends to produce outliers. Although the least-squares criterion is adequate under the Gaussian assumption and usually leads to analytically tractable results, it is no longer appropriate for an impulsive non-Gaussian environment, largely due to its lack of robustness against outliers [Bloomfield and Steiger, 1983; Gonin and Money, 1989]. It has been demonstrated many times in the literature that least-squares estimates can deteriorate dramatically when only a small proportion of extreme observations is present in the data [Tarantola, 1987].

The absence of a finite variance does not mean, however, that there are no other adequate measures of variability of stable random variables. As it will be shown later in the book, the *dispersion* of a stable random variable plays a role analogous to the *variance*. For example, the larger the dispersion of a stable distribution, the more it spreads around its median. Hence, the *minimum dispersion* criterion becomes a natural and mathematically meaningful choice as a measure of optimality in signal processing problems based on stable models. By minimizing the error dispersion, we minimize the average magnitude of estimation errors. Furthermore, it has been shown that minimizing the dispersion is also equivalent to minimizing the probability of large estimation errors [Cline and Brockwell, 1985]. Hence, the minimum dispersion criterion is well justified under the stable model assumption. It is a direct generalization of the minimum mean-squared error criterion (they are the same in the Gaussian case) and relatively simple to calculate.

Minimizing the dispersion is also equivalent to minimizing the fractional lower-order moments of estimation errors that measure the L_p distance between an estimate and its true value, for $0 < p < \alpha \leq 2$. This result is not surprising since the L_p norms for $p < 2$ are well known for their robustness against outliers such as those that may be described by the stable law [Claerbout and Muir, 1973; Schroeder and Yarlagadda, 1989; Yarlagadda, Bednar, and Watt, 1985; Schroeder, 1991; Taylor, Banks, and Mccoy, 1979]. It is also known that all the lower-order moments of a stable random variable are equivalent, i.e., any two of the lower-order moments differ by a fixed constant that is independent of the random variable itself. A common choice is the L_1 norm, which is sometimes very convenient.

Stable signal processing based on fractional lower-order moments will inevitably introduce nonlinearity to even linear problems. The basic reason for the nonlinearity is that we have to solve linear estimation problems in Banach or metric spaces instead of Hilbert spaces. It is well known that, while the linear space generated by a Gaussian process is a Hilbert space, the linear space of a stable process is a Banach space when $1 \leq \alpha < 2$ and only a metric space when $0 < \alpha < 1$ [Cambanis and Miller, 1981]. Banach or metric spaces do not have as nice properties and structures as Hilbert spaces for linear estimation problems. For example, while any finite number of Gaussian random variables can be expressed as linear combinations of *independent* Gaussian random variables, it is shown by Schilder [1970] that the representation of even two stable random variables of the same characteristic exponent as linear combinations of finitely many independent stable random variables is impossible.

Despite the aforementioned difficulties, significant progress has been made in developing a linear estimation theory for stable processes over the past thirty years. In this book, we will give an overview of some of the results that are relevant to digital signal processing problems.

1.4 APPLICATIONS OF STABLE DISTRIBUTIONS AND STABLE SIGNAL PROCESSING

The concept of stable distributions was first introduced by Lévy [1925] in the study of generalized central limit theorems. They are direct generalizations of Gaussian distributions and share a lot of useful properties of Gaussian laws. Yet, despite all the attractive properties, the stable laws have attracted little attention from researchers in signal processing [Stuck and Kleiner, 1974; Stuck, 1978; Cambanis and Miller, 1981]. There are basically two reasons for this. First, stable laws do not have explicit expressions for their densities or distributions except for the Gaussian ($\alpha = 2$) and Cauchy ($\alpha = 1$) distributions. But power series expansions do exist for the probability densities functions. With today's computational power, numerical integrations are inexpensive to carry out. Moreover, much of the work that ordinarily uses probability density functions can be carried out in the transform domain of characteristic functions. The second main reason for the unpopularity of stable laws among engineers is apparently due to the fact that the pth moments of a $S\alpha S$ random variable exist only for $p < \alpha$ [Feller, 1966]. Thus for all non-Gaussian stable distributions there are no finite second-order moments and for $\alpha \leq 1$ not even finite first-order moments exist. Since the second-order moment or the variance is often associated with the concept of power, it seems to be widely felt that infinite variance is inappropriate in almost any signal processing context.

We believe that this kind of reasoning is superficial and inappropriate. This is similar to saying that we should not use irrational numbers at all because we can never have irrationals from any physical measurements. Besides, as it is pointed out by Stuck [1978], the Gaussian distribution would not be a physically appropriate assumption because it is unbounded. The purpose of mathematical modeling is not to account for every single detail of how a physical process is generated but to explain important and relevant features of the process in order to optimally and efficiently extract desired information from the observed data. For example, linear models in system theory are applicable only in a limited range. Beyond this range, nonlinear models must be used. By the same token, Gaussian models may be adequate for modeling a limited range of observed data. For a larger range, an infinite-variance stable model may have to be used [Stuck, 1978]. This is especially true when outliers or heavy tails appear in the observed data. Of course, there are certain situations when neither Gaussian nor non-Gaussian stable distributions may be appropriate.

To illustrate the point, let us look at Figures 1.2 and 1.3. Figure 1.2(a), (c) shows realizations of Gaussian and stable ($\alpha = 1.1$) white noise with zero mean and unit dispersion. Severe impulsiveness of the stable white noise is obvious. Yet, if samples between 30 and 50 are extracted (see (b) and (d)), they do not seem different. Figure 1.3(a)–(d) shows similar conclusions for AR(1) processes with the pole at 0.9.

The most important features of non-Gaussian stable distributions are probabilistic stability and long inverse power tails, in addition to the basic fact that they arise naturally as limit distributions. If a physical phenomenon has both the stability property and long tails, stable distributions could provide useful models. Stability is a natural assumption. A lot of physical processes, such as natural noise sources, have this property. On the other hand, in recent years more and more physical phenomena have been found to have tails that are heavier than the Gaussian [Lerner, 1961; Miller, 1972; Wegman, Schwartz, and Thomas, 1989]. Thus, it can be expected that stable distributions and stable processes will provide useful models

1.4 STABLE DISTRIBUTIONS AND STABLE SIGNAL PROCESSING

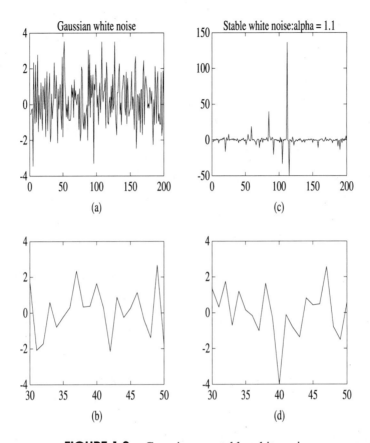

FIGURE 1.2 Gaussian vs. stable white noise.

for many phenomena observed in diverse fields. Indeed, stable laws have found application in physics, economics, hydrology, biology, and electrical engineering.

The earliest application of stable laws was discovered in physics by the Danish astronomer Holtsmark [1919]. He discovered that random fluctuations of gravitational fields of stars in space under certain natural assumptions obey the stable law with characteristic exponent $\alpha = 1.5$. However, the stable distribution did not receive serious attention until the work by Mandelbrot and his followers in economics and finance in the sixties. Because of the failure of the Gaussian assumption and the least-squares criterion in economic time series, Mandelbrot proposed a revolutionary approach based on stable distributions to the problem of price movement [Mandelbrot, 1963]. Many economical variables, such as common stock price changes, fluctuations in speculative prices, and interest rates, have already been shown to have properties that conform closely to those of non-Gaussian stable laws [Fama, 1965; Press, 1968; Teichmoeller, 1971].

The stable distribution has also found applications in signal processing and communications. For example, Mandelbrot and van Ness [1968] used Gaussian and stable fractional stochastic processes to describe long-range dependence aris-

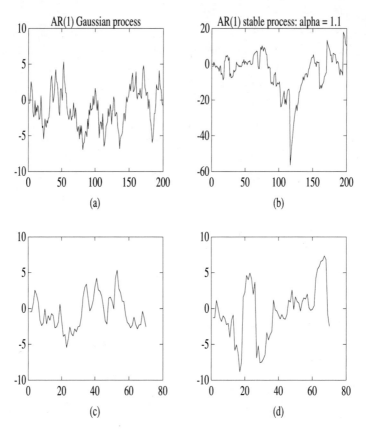

FIGURE 1.3 Gaussian vs. stable AR(1) processes with the pole at 0.9.

ing in engineering, economics, and hydrology. The stable distribution has also been used to describe the patterns of error clustering in telephone circuits [Berger and Mandelbrot, 1963]. However, we feel that the stable distribution deserves more attention, especially in the area of impulsive noise modeling. It is widely acknowledged that, in many communication problems, the conventional additive Gaussian noise assumption is inadequate. This is often due to the occurrence of noise with low probability but large amplitudes. This impulsive component of noise has been found to be significant in many problems, including atmospheric noise and underwater problems such as sonar and submarine communications, where the ambient acoustic noise may include impulses due to ice cracking in arctic regions [Kassam, 1988; Mertz, 1961; Bouvet and Schwartz, 1989]. These types of impulsive noise are often observed to be close to the Gaussian distribution but with heavier tails. For example, it has been reported by Machell [1989] that under-ice and shallow-water noise has a distribution similar to that of Gaussian noise (symmetric, smooth, unimodal, etc.), but has heavier tails. The atmospheric noise may be considered as the result of a large number of independent sources (mainly thunderstorms) in space so that central limit theorems apply. However, the distributions have been shown

[Lerner, 1961; Mertz, 1961; Rappaport and Kurz, 1966] to have algebraic tails, i.e., $x^{-\beta}$ for $1 < \beta < 3$, a characteristic associated with the stable distribution. In Miller [1972], it is proposed to find a useful class of noise distributions with algebraic rather than exponential decay of the density of impulsive non-Gaussian noise that would approach the Gaussian distribution as some parameter approaches some limit. All these evidences suggest the use of stable laws as appropriate models. In fact, the Cauchy distribution, which is a member of the stable family, has been used in several papers, such as Rappaport and Kurz [1966], to represent severe impulsive noise. Stuck and Kleiner [1974] empirically found that the noise over certain telephone lines can be best described by stable laws with characteristic exponent α close to 2. They suggested that the design of receivers should take into account this particular noise characteristic [Berger and Mandelbrot, 1963].

PROBLEMS

1. Consider the signal detection problem:

$$H_1 : r(u) = s + n(u),$$
$$H_0 : r(u) = n(u),$$

where $r(u)$ is the observation, s is the signal, and $n(u)$ is the additive noise. Assume s is a positive constant. Find the test statistic, probability of detection, and probability of false alarm as functions of signal s and noise dispersion γ when $n(u)$ is: (1) white Gaussian noise, (2) Cauchy noise with same dispersion.

2. Find the characteristic function $\varphi(\omega)$ of the Cauchy density function:

$$f(x) = \frac{\gamma}{\pi(\gamma^2 + x^2)}, \quad -\infty < x < \infty.$$

Is $\varphi(\omega)$ differentiable at $\omega = 0$?

3. Let Y be a real random variable. Find the constant c that minimizes $R(c)$, where $R(c)$ is defined as: (1) $R(c) = E\{(Y - c)^2\}$, (2) $R(c) = E\{|Y - c|\}$. In other words, show that: (1) $c = E\{Y\}$, (2) $c = \text{Median}(Y)$.

4. If $T(\mathbf{X})$ is an estimate of the parameter $q(\theta)$ and both functions are real, show that $E\{|T(\mathbf{X}) - q(\theta)|\} \leq \sqrt{E\{(T(\mathbf{X}) - q(\theta))^2\}}$. Equality holds when $T(\mathbf{X})$ is a constant.

5. Optimal predictor problem: given the joint p.d.f. of a real random vector (variable) $\mathbf{X}$ and a real random variable Y, find the predictor $g(\mathbf{X})$ to minimize: (1) the mean square error $E\{(g(\mathbf{X}) - Y)^2\}$, (2) the mean absolute error $E\{|g(\mathbf{X}) - Y|\}$.

6. If (X, Y) has a bivariate normal distribution, show that the conditional mean of Y given X is the same as the conditional median, i.e., $E\{Y|X\} = \text{Median}\{Y|X\}$.

7. This problem deals with some of the mathematical properties of LAD (Least Absolute Deviation) method for curve fitting. Given n points $(\mathbf{x}_i, y_i) \in \mathbf{R}^{k+1}$,

$i = 1, 2, \ldots, n$, where $\mathbf{x}_i$ is $1 \times k$ row vector and y_i is a scalar, the LAD fitting problem (which is essentially a linear programming optimization problem) is to find $\hat{\mathbf{c}} \in \mathbf{R}^k$ to minimize the absolute distance function:

$$f(\mathbf{c}) = \sum_{i=1}^{n} |y_i - \sum_{j=1}^{k} c_j x_{ij}| = \sum_{i=1}^{n} |y_i - <\mathbf{c}, \mathbf{x}_i>| = \|\mathbf{y} - \mathbf{Xc}\|_1,$$

where $\mathbf{X} = (\mathbf{x}_i)$, $\mathbf{x}_i$ is the ith row of matrix $\mathbf{X}$. Prove: (1) Function $f(\cdot)$ is continuous and convex. (2) If $\mathbf{X}$ has a full rank, i.e., rank($\mathbf{X}$) $= k \leq n$, then the solutions exist and they are bounded. Otherwise, the solutions could be unbounded.

8. (Computer project) A linear regression model is described by:

$$y_i = \theta_1 x_{i1} + \theta_2 x_{i2} + \epsilon_i.$$

Estimate θ_1 and θ_2 from n points of $(x_{i1}, x_{i2}, y_i), i = 1, 2, \ldots, n$ by the least square method and the least absolute error method when ϵ_i has: (1) standard normal distribution, (2) standard Cauchy distribution.

(*Hint*: One can refer to Chapter 2 of this book for a general method to generate stable random numbers.)

2

The Stable Distribution

2.1 INTRODUCTION

In this chapter, we introduce the stable distribution and its characterizations and statistical properties. The proofs of most of the results are omitted and can be found in the probability and statistics literature [e.g., Lukacs, 1960; Feller, 1966; Gnedenko and Kolmogorov, 1968; Miller, 1978; Zolotarev, 1986].

2.2 THE CHARACTERISTIC FUNCTION

The stable distribution can be most conveniently described by its characteristic function as follows:

Definition 1 (Name of Definition) *A univariate distribution function $F(x)$ is stable if and only if its characteristic function has the form*

$$\varphi(t) = \exp\{jat - \gamma |t|^\alpha [1 + j\beta \text{sign}(t)\omega(t, \alpha)]\} \quad (2.1)$$

where

$$\omega(t, \alpha) = \begin{cases} \tan \frac{\alpha\pi}{2}, & \text{if } \alpha \neq 1 \\ \frac{2}{\pi} \log |t|, & \text{if } \alpha = 1, \end{cases} \quad (2.2)$$

$$\text{sign}(t) = \begin{cases} 1, & \text{if } t > 0 \\ 0, & \text{if } t = 0 \\ -1, & \text{if } t < 0 \end{cases} \quad (2.3)$$

and

$$-\infty < a < \infty, \; \gamma > 0, \; 0 < \alpha \leq 2, \; -1 \leq \beta \leq 1. \quad (2.4)$$

Thus, a stable characteristic function (or distribution) is completely determined by four parameters: α, a, β, and γ, where:

1. α is called the *characteristic exponent*. It is uniquely determined. The distributions and corresponding random variables are called α-*stable*. The characteristic exponent measures the "thickness" of the tails of the distribution. Thus, if a stable random variable is observed, the larger the value of α, the less likely it is to observe values of the random variable, which are far from its central location. A small value of α will imply considerable probability mass in the tails of the distribution. An $\alpha = 2$ corresponds to a Gaussian distribution (for any β), while $\alpha = 1, \beta = 0$ correspond to a Cauchy distribution.
2. γ is a scale parameter, also called the *dispersion* [Stuck, 1978]. It is similar to the variance of the Gaussian distribution and equals half the variance in that Gaussian case (i.e., when $\alpha = 2$).
3. β is a symmetry parameter. $\beta = 0$ implies a distribution symmetric about a. In this case, the distribution is called *symmetric α-stable*, or simply *SαS*. Symmetric stable distributions represent an important subclass of stable distributions. The Gaussian and the Cauchy distributions are both *SαS*.
4. a is a location parameter. For *SαS* distributions, it is the mean when $1 < \alpha \leq 2$ and the median when $0 < \alpha < 1$.

A stable distribution with parameter α is often called α-*stable* and is said to be *standard* if $a = 0, \gamma = 1$. Figure 2.1 illustrates *SαS* standard characteristic functions. It is easy to show that if X is stable with parameters α, β, γ, and a, then $(X - a)/\gamma^{\frac{1}{\alpha}}$ is a standardized variable with characteristic exponent α and symmetry parameter β. Note that according to this definition, the standard Gaussian distribution has variance 2, not 1.

The parameters characterizing *SαS* distributions are summarized in Table 2.1.

2.3 THE PROBABILITY DENSITY FUNCTION

By taking the inverse Fourier transform of the characteristic function, it is easy to show that the standard stable density function is given by

$$f(x; \alpha, \beta) = \frac{1}{\pi} \int_0^\infty \exp(-t^\alpha) \cos[xt + \beta t^\alpha \omega(t, \alpha)] dt. \qquad (2.5)$$

Note that $f(x; \alpha, \beta) = f(-x; \alpha, -\beta)$. It can also be shown that the probability density functions of stable distributions are bounded and have derivatives of arbitrary orders [Zolotarev, 1986].

Unfortunately, no closed-form expressions exist for the general stable density and distribution functions, except for the Gaussian ($\alpha = 2$), Cauchy ($\alpha = 1, \beta = 0$), and Pearson ($\alpha = \frac{1}{2}, \beta = -1$) distributions [Holt and Crow, 1973]. But power se-

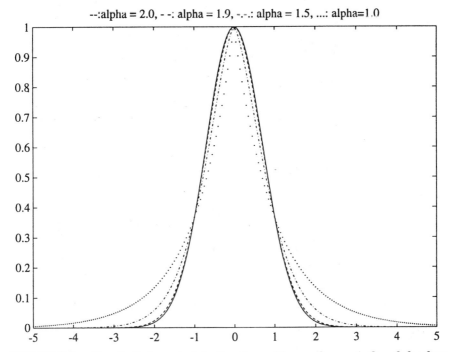

FIGURE 2.1 The SαS characteristic function with $a = 0, \gamma = 1, \beta = 0$ for four different values of the characteristic exponent α.

ries expansions of stable density functions are available. The standard stable density function can be expanded into absolutely convergent series as follows [Bergstrom, 1952; Feller, 1966]: for $x > 0$

$$f(x; \alpha, \beta) = \begin{cases} \dfrac{1}{\pi x} \displaystyle\sum_{k=1}^{\infty} \dfrac{(-1)^{k-1}}{k!} \Gamma(\alpha k + 1) \left(\dfrac{x}{r}\right)^{-\alpha k} \sin\left[\dfrac{k\pi}{2}(\alpha + \zeta)\right], & 0 < \alpha < 1 \\ \dfrac{1}{\pi x} \displaystyle\sum_{k=1}^{\infty} \dfrac{(-1)^{k-1}}{k!} \Gamma\left(\dfrac{k}{\alpha} + 1\right) \left(\dfrac{x}{r}\right)^{k} \sin\left[\dfrac{k\pi}{2\alpha}(\alpha + \zeta)\right], & 1 < \alpha \leq 2 \end{cases}$$

(2.6)

TABLE 2.1 Parameters of the SαS Distribution

	SαS(X)	Gaussian (Y)	Cauchy (Z)
α	Characteristic Exponent	2	1
a	Location Parameter	Mean	Median
γ	Dispersion	Variance/2	Dispersion
Normalization	$(X - a)/\gamma^{1/\alpha}$	$(Y - a)/\sqrt{2\gamma}$	$(Z - a)/\gamma$

where

$$\eta = \beta\tan(\pi\alpha/2), r = (1+\eta^2)^{-1/(2\alpha)}, \zeta = -(2/\pi)\arctan\eta \qquad (2.7)$$

and $\Gamma(\cdot)$ is the usual gamma function defined by

$$\Gamma(x) = \int_0^\infty t^{x-1}e^{-t}dt. \qquad (2.8)$$

In particular, the standard $S\alpha S$ density function is given by

$$f_\alpha(x) = \begin{cases} \dfrac{1}{\pi x}\sum_{k=1}^\infty \dfrac{(-1)^{k-1}}{k!}\Gamma(\alpha k+1)|x|^{-\alpha k}\sin\left[\dfrac{k\alpha\pi}{2}\right], & 0 < \alpha < 1 \\ \dfrac{1}{\pi\alpha}\sum_{k=0}^\infty \dfrac{(-1)^k}{2k!}\Gamma\left(\dfrac{2k+1}{\alpha}\right)x^{2k}, & 1 \leq \alpha \leq 2. \end{cases} \qquad (2.9)$$

For the $S\alpha S$ distributions with $\alpha > 1$, asymptotic series are available for density functions [Bergstrom, 1952]. In particular, for the standard $S\alpha S$ density function one has

$$f_\alpha(x) = \sum_{k=0}^n a_k x^{2k} + O(|x|^{2n+1}), \qquad (2.10)$$

as $|x| \to 0$, and

$$f_\alpha(x) = \sum_{k=1}^n \frac{b_k}{|x|^{\alpha k+1}} + O(|x|^{-\alpha(n+1)-1}), \qquad (2.11)$$

as $|x| \to \infty$.[1] In Eqs. (2.10) and (2.11), we have

$$a_k = \frac{1}{\pi\alpha}\frac{(-1)^k}{(2k)!}\Gamma\left(\frac{2k+1}{\alpha}\right) \qquad (2.12)$$

$$b_k = -\frac{1}{\pi}\frac{(-1)^k}{k!}\Gamma(\alpha k+1)\sin\left(\frac{k\alpha\pi}{2}\right). \qquad (2.13)$$

If the asymptotic series (2.10) or (2.11) are to be computed for a large number n of terms, inaccuracies may arise due to the computation for large argument of the gamma functions appearing in Eqs. (2.12) and (2.13). These difficulties can be

[1] Eq. (2.10) is an asymptotic expansion for all $0 < \alpha < 2$, $|x| \to 0$, and becomes a convergent series for $1 < \alpha < 2$. Eq. (2.11) is an asymptotic expansion for all $0 < \alpha < 2$, $|x| \to \infty$, and becomes a convergent series for $0 < \alpha < 1$.

minimized, however, by the following procedure. We observe that:

$$a_k x^{2k} = -[a_{k-1} x^{2(k-1)}] \frac{\Gamma(\frac{2k+1}{\alpha})}{\Gamma(\frac{2k-1}{\alpha})} \frac{x^2}{2k(2k-1)}, \quad k = 1, 2, 3, \ldots \quad (2.14)$$

with

$$a_0 = \frac{1}{\pi \alpha} \Gamma\left(\frac{1}{\alpha}\right). \quad (2.15)$$

For large k, we have [Abramowitz and Stegun, 1965]

$$\frac{\Gamma(\frac{2k+1}{\alpha})}{\Gamma(\frac{2k-1}{\alpha})} \approx \left(\frac{2k}{\alpha}\right)^{\frac{2}{\alpha}} \left[1 - \frac{1}{2k} + \frac{1}{3} \frac{\alpha + 2}{(2k)^2}\right], \quad (2.16)$$

which, if substituted back in Eq. (2.14), gives a recursive formula for the accurate evaluation of the asymptotic series (2.10). Similarly, we use the formulae

$$\frac{b_k}{|x|^{\alpha k + 1}} = -\left[\frac{b_{k-1}}{|x|^{\alpha(k-1)+1}}\right] \frac{(\alpha k + 1)\alpha}{k|x|^\alpha} \frac{\sin(\frac{k\pi\alpha}{2})}{\sin(\frac{(k-1)\pi\alpha}{2})}$$

$$\cdot [1 - \frac{\alpha(\alpha+1)}{2} \frac{1}{\alpha k + 1} + \frac{1}{24} \frac{\alpha(\alpha-1)(\alpha+1)(3\alpha+4)}{(\alpha k + 1)^2}],$$

for large k, with $\qquad (2.17)$

$$\frac{b_1}{|x|^{\alpha+1}} = \frac{1}{\pi} \frac{\Gamma(\alpha+1)\sin(\frac{\alpha\pi}{2})}{|x|^{\alpha+1}}, \quad (2.18)$$

to recursively compute the asymptotic series (2.11).

The standard $S\alpha S$ density functions for different values of the characteristic exponent α are shown in Figure 2.2, with $\alpha = 2$ corresponding to the Gaussian density with zero mean and variance equal to 2 and $\alpha = 1$ corresponding to the Cauchy density. Observe that $S\alpha S$ densities maintain many of the features of the Gaussian density. They are smooth, unimodal, symmetric with respect to the median, and bell-shaped. A detailed comparison between the standard normal and $S\alpha S$ density functions shows that non-Gaussian stable density functions differ from the corresponding Gaussian density in the following ways. For small absolute values of x, the $S\alpha S$ densities are more peaked than the normal. For some intermediate range of $|x|$, the $S\alpha S$ distributions have lower densities than the normal. Most importantly, unlike the Gaussian density, which has exponential tails, the stable densities have algebraic tails [Gnedenko and Kolmogorov, 1968]. Thus $S\alpha S$ densities have heavier tails than the Gaussian. The smaller α is, the heavier are the tails. This is a desir-

18 THE STABLE DISTRIBUTION

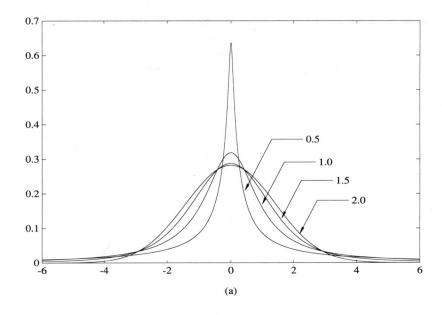

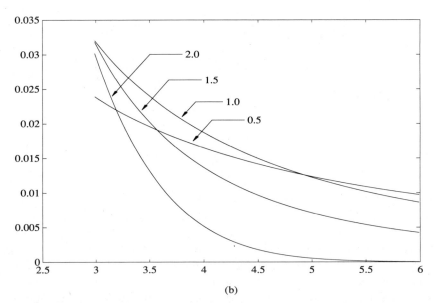

FIGURE 2.2 Density functions of $S\alpha S$ distributions for different values of the characteristic exponent α: (a) the overall densities and (b) the tails of the densities.

2.4 AN EFFICIENT METHOD FOR COMPUTING $S\alpha S$ DENSITY FUNCTIONS

able feature for many applications in signal processing since many non-Gaussian phenomena are similar to the Gaussian phenomenon, but with heavier tails [Mertz, 1961; Lerner, 1961; Machell, Penrod, and Ellis, 1989; Rappaport and Kurz, 1966].

2.4 AN EFFICIENT METHOD FOR COMPUTING $S\alpha S$ DENSITY FUNCTIONS

The curves in Figure 2.2 have been produced by calculation of the inverse Fourier transform integral in Eq. (2.5). This procedure is quite accurate for computation of $S\alpha S$ p.d.f.'s; however, it is not readily applicable for real-time calculations because of the extensive numerical integrations it requires. Therefore, alternative expressions need to be used for real-time computation of a $S\alpha S$ p.d.f. As a first step, we use the asymptotic series in Eqs. (2.10) and (2.11). In particular, Eq. (2.10) is expected to give a good approximation to a $S\alpha S$ p.d.f. $f_\alpha(x)$ as long as its argument is small, while Eq. (2.11) will give a good approximation to $f_\alpha(x)$ for large argument. We have found, however, that there exists an interval of values of the argument of $f_\alpha(x)$ for which none of the asymptotic series in Eqs. (2.10) and (2.11) is sufficiently accurate. This is clearly illustrated in Figure 2.3, where we show plots of the $S(\alpha = 1.5)S$ p.d.f. (with $\delta = 0$ and $\gamma = 1$) obtained with the following three different methods: (i) by numerical computation of the inverse Fourier transform

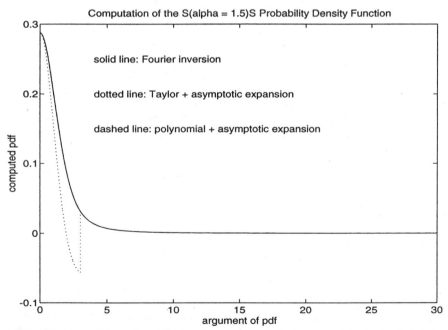

FIGURE 2.3 Comparison of the three methods for computing the $S\alpha S$ density function for $\alpha = 1.5$.

in Eq. (2.5), (ii) via the asymptotic series expansions in Eqs. (2.10) and (2.11), in which the first $n = 100$ terms of the Taylor series expansion have been computed for $|\xi| \leq 3$ and the first $n = 9$ terms of the asymptotic expansion have been computed for $|x| > 3$, and (iii) by the method described in the next paragraph. Clearly, the asymptotic series approximation of Eq. (2.10) is accurate only for very small values of the argument of the p.d.f. and deteriorates very fast for larger values. On the other hand, the asymptotic series of Eq. (2.11) is not accurate for small arguments for which the series in Eq.(2.10) is not valid.

For the real-time computation of $S\alpha S$ p.d.f.'s at arbitrary argument, a method was devised by Tsihrintzis and Nikias [1995b] and found very accurate and efficient to implement. The method relies on the fact [Zolotarev, 1986] that the $S\alpha S$ p.d.f.'s are entire analytic functions, thus having bounded derivatives of all orders. Our approach, therefore, for the real-time computation of a $S\alpha S$ p.d.f. f_α consists of three steps: (i) choice of a cutoff argument, such that the p.d.f. at larger values can be accurately computed via the first few terms of Eq. (2.11), (ii) evaluation of the p.d.f. at a number of points in the interval (0, cutoff) via numerical computation of the Fourier integral in Eq. (2.5), and (iii) evaluation of the coefficients of an interpolating polynomial of small degree from the values of the p.d.f. at the selected points of the previous step. This method is, indeed, very accurate and is illustrated in Figure 2.3, for the computation of the $S(\alpha = 1.5)S$ p.d.f with $\gamma = 1$ and $\delta = 0$. We have performed this procedure for the $S\alpha S$ p.d.f.'s with $\gamma = 1$, $\delta = 0$, and $\alpha = 0.5, 1.5$, and 1.99 and interpolating polynomials of appropriate degree and asymptotic series of appropriate number of terms. The resulting coefficients in the asymptotic series and in the interpolating polynomials are shown in Table 2.2.

2.5 PROPERTIES OF UNIVARIATE STABLE LAWS

Two of the most important properties of the stable distribution are the *stability property* and the *Generalized Central Limit Theorem*. They are responsible for much of the appeal that stable distributions have as statistical models of uncertainty.

The stability property is actually a defining characteristic of the stable distribution. That is:

Theorem 1 (Stability Property) *A random variable X has a stable distribution if and only if for all X_1, X_2 independent, with the same distribution as X, and for arbitrary constants a_1, a_2, there are constants a and b such that*

$$a_1 X_1 + a_2 X_2 \stackrel{d}{=} aX + b \tag{2.19}$$

where the notation $X \stackrel{d}{=} Y$ means that X and Y have the same distribution. By using the characteristic function of the stable distribution, one can easily show a more general statement: If $X_1, X_2, \ldots, X_n$ are independent and follow a stable law with the same (α, β), then all linear combinations of the form $\sum_{j=1}^{n} a_j X_j$ are stable with the same parameters α and β.

TABLE 2.2 Computation of SαS p.d.f.'s in Real Time

$\alpha =$	$f_\alpha(x) =$
0.5	$-0.3238x^5 + 1.7166x^4 - 3.6100x^3 + 3.8470x^2 - 2.2125x + 0.6709$ if $\|x\| \leq 1.5$ $\dfrac{0.0062}{x^{3.5}} - \dfrac{9.522 \cdot 10^{-11}}{x^3} + \dfrac{0.0499}{x^{2.5}} - \dfrac{0.1592}{x^2} + \dfrac{0.1995}{x^{1.5}}$ if $\|x\| > 1.5$
1.5	$0.0001x^9 - 0.0008x^8 + 0.0048x^7 - 0.0125x^6 + 0.0083x^5$ $+ 0.0196x^4 + 0.0007x^3 - 0.1058x^2 + 0.0016x + 0.2874$ if $\|x\| \leq 3$ $\dfrac{1.4323 \cdot 10^4}{x^{14.5}} + \dfrac{8.1449 \cdot 10^{-5}}{x^{13}} - \dfrac{0.0531 \cdot 10^4}{x^{11.5}} - \dfrac{0.0160 \cdot 10^4}{x^{10}} - \dfrac{0.0026 \cdot 10^4}{x^{8.5}}$ $- \dfrac{1.0284 \cdot 10^{-7}}{x^7} + \dfrac{0.0002 \cdot 10^4}{x^{5.5}} + \dfrac{0.0001 \cdot 10^4}{x^4} + \dfrac{8.1449 \cdot 10^{-5}}{x^{2.5}}$ if $\|x\| > 3$
1.99	$1.442 \cdot 10^{-8} x^{11} - 8.171 \cdot 10^{-7} x^{10} + 1.778 \cdot 10^{-5} x^9 - 0.0002x^8 + 0.0013x^7$ $- 0.0046x^6 + 0.0070x^5 + 0.0011x^4 + 0.0049x^3 - 0.0725x^2 + 0.0014x$ $+ 0.2821$ if $\|x\| \leq 6$ $\dfrac{1.0944 \cdot 10^{12}}{x^{22.89}} + \dfrac{0.0247 \cdot 10^{12}}{x^{20.9}} + \dfrac{0.0006 \cdot 10^{12}}{x^{18.91}} + \dfrac{1.6546 \cdot 10^7}{x^{16.92}} + \dfrac{5.0112 \cdot 10^5}{x^{14.93}} + \dfrac{1.7129 \cdot 10^4}{x^{12.94}}$ $+ \dfrac{671.5207}{x^{10.95}} + \dfrac{30.8253}{x^{8.96}} + \dfrac{1.7011}{x^{6.97}} + \dfrac{0.1164}{x^{4.98}} + \dfrac{0.0099}{x^{2.99}}$ if $\|x\| > 6.$

As a consequence of the stability property, it can be shown that stable distributions are the only possible limit distributions for sums of i.i.d. random variables. This is known as the Generalized Central Limit Theorem and is formally stated as follows [Breiman, 1968]:

Theorem 2 (Generalized Central Limit Theorem) *X is the limit in distribution of normalized sums of the form*

$$S_n = (X_1 + \cdots + X_n)/a_n - b_n$$

where $X_1, X_2, \ldots, X_n$ are i.i.d. and $n \to \infty$, if and only if the distribution of X is stable.

In particular, if the X_i's are i.i.d. and have finite variance, then the limit distribution is Gaussian. This is of course the result of the ordinary Central Limit Theorem.

The main cause of the different behaviors of the Gaussian and the (non-Gaussian) stable distributions is their tails. It can be shown [Lévy, 1925; Weron, 1983] that for a non-Gaussian α-stable random variable X with zero location parameter and dispersion γ:

$$\lim_{t \to \infty} t^\alpha P(|X| > t) = \gamma C(\alpha) \tag{2.20}$$

where $C(\alpha)$ is a positive constant depending on α. Thus, stable laws have inverse power (i.e., algebraic) tails. In contrast, normal distributions have exponential tails. This implies that the tails of stable laws are significantly thicker than those of the normal distribution and, in fact, the smaller the value of α, the thicker the tails.

An important consequence of (2.20) is the *nonexistence* of the second-order moment of stable distributions, except for the limiting case $\alpha = 2$. Specifically, one has the following result.

Theorem 3 *Let X be an α-stable random variable. If $0 < \alpha < 2$, then*

$$E|X|^\delta = \infty \quad \text{if } \delta \geq \alpha$$

and

$$E|X|^\delta < \infty \quad \text{if } 0 \leq \delta < \alpha.$$

If $\alpha = 2$, then

$$E|X|^\delta < \infty \quad \text{for all } \delta \geq 0.$$

Proof: The result for $\alpha = 2$ is well-known. Let us now prove the case where $0 < \alpha < 2$. For a nonnegative random variable Y, it is easy to show by interchanging integration order that

$$\mathbf{E}(Y) = \int_0^\infty P(Y > t)\,dt.$$

Replacing Y by $|X|^p$, one has

$$\begin{aligned}\mathbf{E}(|X|^p) &= \int_0^\infty P(|X|^p > t)\,dt \\ &= \int_0^\infty pu^{p-1}P(|X| > u)\,du.\end{aligned} \qquad (2.21)$$

Since

$$u^{p-1}P(|X| > u) = O(u^{p-1}), \quad \text{as } u \to 0$$

and

$$u^{p-1}P(|X| > u) = O(u^{p-\alpha-1}), \quad \text{as } u \to \infty,$$

one concludes that $\mathbf{E}(|X|^p) < \infty$ if and only if $0 \le p < \alpha$.

Hence for $0 < \alpha \le 1$, α-stable distributions have no finite first- or higher-order moments; for $1 < \alpha < 2$, they have the first moment and all the fractional lower-order moments (FLOMs) of order p where $p < \alpha$; for $\alpha = 2$, all moments exist. In particular, *all non-Gaussian stable distributions have infinite variance.* □

Table 2.3 gives a few FLOMs of the standardized stable distribution for four different values of α. In Chapter 3, we give a closed-form formula for computing the FLOMs.

TABLE 2.3 Absolute pth-Order Moments of the Standardized SαS Distributions

p	$\alpha = 1$(Cauchy)	$\alpha = 1.5$	$\alpha = 1.9$	$\alpha = 2.0$(Gaussian)
0.5	1.4142	1.0804	0.9920	0.9770
1.0	∞	1.7055	1.1903	$\sqrt{\pi}$
1.2	∞	2.5521	1.3421	1.2331
1.5	∞	∞	1.7347	1.4464
1.8	∞	∞	3.3847	1.7431
2.0	∞	∞	∞	2

2.6 MULTIVARIATE STABLE DISTRIBUTIONS

The multivariate stable distribution can again be defined by the stability property:

Definition 2 *A k-dimensional distribution function $F(\mathbf{x}), \mathbf{x} \in R^k$ is called* stable *if, for any independent and identically distributed random vectors $\mathbf{X}_1, \mathbf{X}_2$ with distribution function $F(\mathbf{x})$ and arbitrary constants a_1, a_2, there exist $a \in R, \mathbf{b} \in R^k$ and a random vector $\mathbf{X}$ with the same distribution function $F(\mathbf{x})$ such that*

$$a_1 \mathbf{X}_1 + a_2 \mathbf{X}_2 \stackrel{d}{=} a\mathbf{X} + \mathbf{b}. \tag{2.22}$$

One of the main differences between univariate and multivariate stable distributions is that the family of one-dimensional stable distributions forms a parametric set while the family of multivariate stable distributions forms a nonparametric set, as we can see from its characteristic function [Lévy, 1925; Paulauskas, 1976]:

Theorem 4 *A k-dimensional distribution function $F(\mathbf{x}), \mathbf{x} \in R^k$ is stable if and only if its characteristic function has the following form:*

$$\varphi(\mathbf{t}) = \begin{cases} \exp\{j\mathbf{t}^T \mathbf{a} - \mathbf{t}^T A \mathbf{t}\}, & \text{if } \alpha = 2 \\ \exp\{j\mathbf{t}^T \mathbf{a} - \int_S |\mathbf{t}^T \mathbf{s}|^\alpha \mu(d\mathbf{s}) + j\beta_\alpha(t)\}, & \text{if } 0 < \alpha < 2 \end{cases} \tag{2.23}$$

where

$$\beta_\alpha(\mathbf{t}) = \begin{cases} \tan \frac{\alpha\pi}{2} \int_S |\mathbf{t}^T \mathbf{s}|^\alpha \operatorname{sign}(\mathbf{t}^T \mathbf{s}) \mu(d\mathbf{s}), & \text{if } \alpha \neq 1, 0 < \alpha < 2 \\ \int_S \mathbf{t}^T \mathbf{s} \log |\mathbf{t}^T \mathbf{s}| \mu(d\mathbf{s}), & \text{if } \alpha = 1 \end{cases} \tag{2.24}$$

and where $\mathbf{a}, \mathbf{t} \in R^k$. S is the k-dimensional unit sphere. $\mu(\cdot)$ is a finite Borel measure on S and A is a positive semi-definite symmetric matrix.

Except for the case $\alpha = 2$, multivariate stable distributions form a nonparametric set. They are determined by a vector $\mathbf{a} \in R^k$, a scalar $0 < \alpha < 2$, and a certain finite measure $\mu(d\mathbf{s})$ on the sphere S.

α is the *characteristic exponent* of the stable distribution. It is uniquely determined. If $\alpha = 2$, then the stable distribution is reduced to the multivariate Gaussian distribution with mean $\mathbf{a}$ and covariance matrix $2A$. $\mathbf{a}$ is a location vector. $\mu(\cdot)$ is called the *spectral measure*. For $1 < \alpha \leq 2$, μ is uniquely determined. $\beta_\alpha(t)$ is called the *skewness function*. If $\beta_\alpha(t) \equiv 0$ then such stable distributions are *symmetric* and called symmetric α stable (SαS).

Although, in theory, multivariate stable distributions are absolutely continuous and have continuously differentiable densities, no closed-form expressions exist for the density functions [Paulauskas, 1976]. Like multivariate normal distributions, marginal distributions of a multivariate stable (or SαS) law are all stable (or SαS) with the same characteristic exponent.

2.6 MULTIVARIATE STABLE DISTRIBUTIONS

If $1 < \alpha \leq 2$ then a random vector **X** follows a multivariate stable (or $S\alpha S$) law with characteristic component α if and only if every linear combination of the components of **X** follows a univariate stable (or $S\alpha S$) law with characteristic exponent α [Weron, 1983]. This defining property is well known for the multivariate Gaussian distribution.

The moment property of multivariate stable distributions basically follows that of univariate stable distributions. If $X_1, \ldots, X_n$ are independent and α-stable, then

$$\mathbf{E}(|X_1|^{p_1} \cdots |X_n|^{p_n}) < \infty$$

if and only if $p_i < \alpha, i = 1, \ldots, n$. If $X_1, \ldots, X_n$ are dependent and jointly $S\alpha S$, then

$$\mathbf{E}(|X_1|^{p_1} \cdots |X_n|^{p_n}) < \infty \qquad (2.25)$$

if and only if

$$0 < p_1 + \cdots + p_n < \alpha,$$

provided that $X_1, \ldots, X_n$ are *n-fold dependent*. This condition is very weak and is almost always satisfied in practice. For details, see Miller [1978].

Although the multivariate Gaussian and stable distributions are similar in many ways, they also differ differently. For example, it is shown in most engineering probability textbooks that any Gaussian random vector can be whitened. Specifically, if **X** is a Gaussian vector, then **X** can be written as

$$\mathbf{X} = A\mathbf{Y}$$

where A is a constant matrix and **Y** is a Gaussian random vector with independent components. But in the stable case, representation of even two stable variables with characteristic exponent α, $0 < \alpha < 2$, as the linear combination of a finite number of independent stable variables of the same characteristic exponent is generally impossible [Schilder, 1970]. This remarkable result suggests that we have to be very careful when we generalize results about Gaussian random variables to non-Gaussian stable random variables.

The characteristic function of the form

$$\varphi(\mathbf{t}) = \exp\left(-\left(\frac{1}{2}(\mathbf{t}^T R \mathbf{t})^{\alpha/2}\right)\right), \qquad (2.26)$$

where the matrix R is positive definite, defines an important subclass of multivariate $S\alpha S$ random vectors, the so-called α-*sub-Gaussian* random vectors [Cambanis and Miller, 1981]. This subclass is often denoted by α-SG(R). It is known [Cambanis

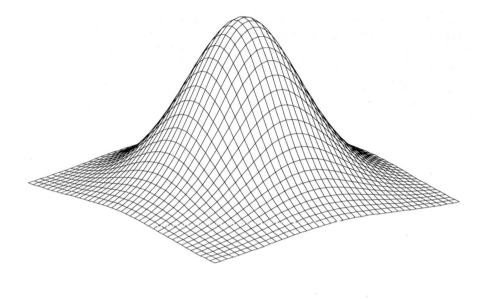

(a)

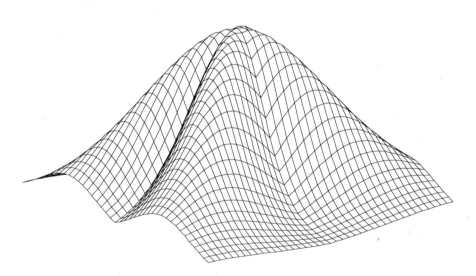

(b)

FIGURE 2.4 α-sub-Gaussian densities with $\alpha = 1.5$ and for two choices of R.

and Miller, 1981] that if $\mathbf{X} \in \alpha\text{-SG(R)}$, then

$$\mathbf{X} = \eta^{\frac{1}{2}}\mathbf{Y}, \qquad (2.27)$$

where η is a positive $\frac{\alpha}{2}$-stable random variable and $\mathbf{Y}$ is a Gaussian vector with mean zero and covariance matrix R. In addition, η and $\mathbf{Y}$ are independent.

Example 2.1: Consider an α-sub-Gaussian with $\alpha = 1.5$. Its density function is plotted in Figure 2.4 for two cases of R:
First case:

$$R = \begin{bmatrix} 1 & 0 \\ 0 & 1 \end{bmatrix}.$$

Second case:

$$R = \begin{bmatrix} 1 & 1 \\ 1 & 2 \end{bmatrix}.$$

□

2.7 EXPONENTIAL CORRELATION FUNCTIONS

The class of symmetric stable characteristic functions coincides with that of (normalized) exponential correlation functions defined by

$$R(\tau) = \exp(-\gamma|\tau|^{\alpha}), \quad 0 < \alpha \leq 2, \quad \gamma > 0.$$

The SαS density functions thus correspond to the power spectral density functions defined by the exponential correlation functions $R(\tau)$. This implies that the results of numerous investigations concerning stable probability distributions can be directly applied to the study of power spectral densities and spectral distributions corresponding to exponential correlations functions.

2.8 GENERATION OF STABLE RANDOM VARIATES

Chambers, Mallows, and Stuck [1976] presented an accurate and inexpensive algorithm for simulating stable random variables for arbitrary characteristic exponent $\alpha(0 < \alpha \leq 2)$ and skewness parameter $\beta(-1 \leq \beta \leq 1)$. The algorithm involves a nonlinear transformation of two independent uniform random variables into one stable random variable. This stable random variable is a continuous function of each of the uniform random variables and of α and a modified skewness parameter β' throughout their respective permissible ranges.

Specifically, suppose we want to generate a random sample X from the standard (α, β) stable distribution, where $0 < \alpha \le 2, -1 \le \beta \le 1$. If $\alpha = 1$, define

$$\beta_A = \beta, \quad \gamma_A = \pi/2,$$

and if $\alpha \ne 1$, define

$$k(\alpha) = 1 - |1 - \alpha|$$
$$\beta_A = 2\arctan(\beta/\cot(\pi\alpha/2))/(\pi k(\alpha))$$
$$\gamma_B = \cos(\pi\beta_A k(\alpha)/2)$$
$$\Phi_0 = -0.5\pi\beta_A(k(\alpha)/\alpha).$$

Furthermore, define

$$\beta' = \begin{cases} -\tan(0.5\pi(1-\alpha))\tan(\alpha\Phi_0), & \alpha \ne 1 \\ \beta_A, & \alpha = 1. \end{cases}$$

Then, $Y = X/\gamma_B^{\frac{1}{\alpha}}$ has the following characteristic function:

$$\varphi_Y(t) = \begin{cases} \exp(-|t|^\alpha - jt(1-|t|^{\alpha-1})\beta'\tan(0.5\pi\alpha)), & \alpha \ne 1 \\ \exp(-|t|(1 + \frac{2}{\pi}j\beta'\log|t|\operatorname{sign}(t))), & \alpha = 1. \end{cases}$$

Now, the random variable Y can be generated as follows. First, generate two independent samples Φ and W, where Φ is uniform on $(-\frac{1}{2}\pi, \frac{1}{2}\pi)$ and W is exponentially distributed with unit mean. Now define the following quantities:

$$\epsilon = 1 - \alpha,$$
$$\tau = -\epsilon \tan(\alpha\Phi_0),$$
$$a = \tan(0.5\Phi),$$
$$B = \tan(0.5\epsilon\Phi)/(0.5\epsilon\Phi),$$
$$b = \tan(0.5\epsilon\Phi),$$
$$z = \frac{\cos(\epsilon\Phi) - \tan(\alpha\Phi_0)\sin(\epsilon\Phi)}{W\cos\Phi},$$
$$d = \frac{z^{\epsilon/\alpha} - 1}{\epsilon}.$$

Finally,

$$Y = \frac{2(a-b)(1+ab) - \Phi\tau B(b(1-a^2) - 2a)}{((1-a^2)(1+b^2))}(1+\epsilon d) + \tau d.$$

Example 2.2: Consider the $S\alpha S$ distribution with $\alpha = 1.5$. In this case, $\beta = 0$, hence $\beta_A = 0, \gamma_B = 1, \Phi_0 = 0$ and $X = Y$. Now suppose $\Phi = 0.3, W = 1.2$, then

$$\epsilon = 1 - \alpha = -0.5,$$

$$\tau = -\epsilon \tan(\alpha \Phi_0) = 0,$$

$$a = \tan(0.5\Phi) = 0.1511,$$

$$B = \tan(0.5\epsilon\Phi)/(0.5\epsilon\Phi) = 1.0019,$$

$$b = \tan(0.5\epsilon\Phi) = -0.0751,$$

$$z = \frac{\cos(\epsilon\Phi) - \tan(\alpha\Phi_0)\sin(\epsilon\Phi)}{W \cos \Phi} = 0.8625,$$

$$d = \frac{z^{\epsilon/\alpha} - 1}{\epsilon} = -0.1011.$$

Finally,

$$X = Y = \frac{2(a-b)(1+ab) - \Phi\tau B(b(1-a^2) - 2a)}{((1-a^2)(1+b^2))}(1 + \epsilon d) + \tau d = 0.4782.$$

□

For the details of the algorithm, readers are referred to the original paper [Chambers, Mallows, and Stuck, 1976]. The Fortran routine written by Chambers, Mallows and Stuck [1976] is also included in the IMSL Library under the name RNSTA.

2.9 CONCLUSION

In this chapter we have introduced the definitions and properties of the stable distribution. It has been shown that the stable distribution enjoys many useful properties that have made the Gaussian distribution popular, such as the linear stability property and the Generalized Central Limit Theorem. One of the main characteristics of the stable distribution is that its density function has an inverse power (algebraic) tail. As we demonstrate later in this book, the algebraic nature of the tails serves as the basis of using the stable distribution to model impulsive noise.

The concept of the stable distribution was introduced by Lévy [1925] about seventy years ago. Since then, many remarkable properties about the stable distribution have been discovered, some of which are contained in a number of well-known probability textbooks. The well-known survey article of Holt and Crow [1973] summarizes basic properties and provides tables and graphs of the stable distribution functions. The first monograph on the stable distribution was published in 1986 written by Zolotarev. It discusses, in detail, analytic properties of stable distributions. Finally, the first engineering tutorial paper on stable distributions, emphasizing digital signal processing applications, was written by Shao and Nikias [1993a].

PROBLEMS

1. The p.d.f. $f(x)$ of a stable distribution has been given in terms of infinite power series (Eq. (2.5) in Sec. 2.3). Prove: (1) $f(x) \geq 0$, and (2) $\int_{-\infty}^{\infty} f(x)\,dx = 1$, i.e., it is indeed a p.d.f.

2. Gaussian, Cauchy, and Pearson distributions are three stable distributions known to have closed-form p.d.f.'s. Derive the p.d.f. from their characteristic functions of the aforementioned stable distributions and find their mean and variance, if they exist. (1) Normal ($\alpha = 2$, $\beta = 0$), (2) Cauchy ($\alpha = 1$, $\beta = 0$), (3) Pearson ($\alpha = 0.5$, $\beta = -1$).

3. Suppose X and X_1, X_2 are i.i.d. with stable distribution F, which has parameters a, γ, α, β. Then for any $p_1 > 0$, $p_2 > 0$, find $p > 0$ and q, such that

$$p_1 X_1 + p_2 X_2 \stackrel{d}{=} pX + q.$$

4. Suppose X and $X_1, X_2, \ldots, X_n$ are i.i.d. with a common $S\alpha S$ distribution F. Prove that for any positive constants $c_1, c_2, \ldots, c_n$, we have

$$\sum_{k=1}^{n} c_k X_k \stackrel{d}{=} \left(\sum_{k=1}^{n} c_k^{\alpha} \right)^{1/\alpha} X.$$

5. Given $X = \sum_{k=1}^{n} c_k X_k$, where X_k ($k = 1, 2, \ldots, n$) are stable r.v.'s with common α, individual skewness β_k, scale parameter γ_k, and location parameter a_k, and where c_k are arbitrary real number, find the parameters α, β, γ, a for X in terms of those of X_k. (Hint: Consider both $\alpha \neq 1$ and $\alpha = 1$.)

6. Show that the Gaussian distribution has an exponential tail while the Cauchy distribution has an algebraic tail.

7. Show that the non-Gaussian $S\alpha S$ processes have algebraic tails. Specifically, show that for $\alpha < 2$, there exists a constant K, such that $1 - F(t) \to Kt^{-\alpha}$ as $t \to \infty$, where $F(t)$ is the $S\alpha S$ distribution function. Further prove that $\lim_{t \to \infty} t^{\alpha} \Pr(|X| > t) = \gamma C(\alpha)$, where γ is the dispersion and $C(\alpha)$ is a positive constant depending on α.

8. If $F(x; \alpha, \beta)$ is a stable distribution, prove that

$$\lim_{x \to \infty} \frac{1 - F(x; \alpha, \beta) - F(-x; \alpha, \beta)}{1 - F(x; \alpha, \beta) + F(-x; \alpha, \beta)} = \beta.$$

9. X is a random variable with stable p.d.f. $f(x)$ and c.f. $\varphi(\omega)$. For what real value of δ is $E\{|X|^{\delta}\}$ infinite? Prove your claim.

3

Symmetric Stable Random Processes

3.1 INTRODUCTION

Although skewed stable distributions are important in certain applications [Mandelbrot, 1963], we will focus our attention on symmetric α-stable distributions in this book.

Recall that a real random variable (r.v.) X is $S\alpha S$, if its characteristic function is of the form

$$\varphi(t) = \exp\{jat - \gamma|t|^\alpha\} \tag{3.1}$$

where $0 < \alpha \leq 2$ is the characteristic exponent, $\gamma > 0$ the dispersion, and $-\infty < a < \infty$ the location parameter. Obviously, when $\alpha = 2$, X is Gaussian, and when $\alpha = 1$, X is Cauchy.

The real r.v.'s $X_1, \ldots, X_n$ are jointly $S\alpha S$, or the real random vector $\mathbf{X} = (X_1, \ldots, X_n)^T$ is $S\alpha S$, if their joint characteristic function is of the form

$$\varphi(\mathbf{t}) = \exp\left\{j\mathbf{t}^T\mathbf{a} - \int_S |\mathbf{t}^T\mathbf{s}|^\alpha \, \mu(d\mathbf{s})\right\} \tag{3.2}$$

where the spectral measure $\mu(\cdot)$ is symmetric, i.e., $\mu(A) = \mu(-A)$ for any measurable set A on the unit sphere S, and $\mathbf{a}, \mathbf{t}$, and $\mathbf{s} \in R^n$. When $1 < \alpha \leq 2$, the real r.v.'s $X_1, \ldots, X_n$ are jointly $S\alpha S$, if and only if all the linear combinations $a_1X_1 + a_2X_2 + \cdots + a_nX_n$ are $S\alpha S$.

A collection of random variables $\{X(t), t \in T\}$, where T is an arbitrary index set, is said to be a $S\alpha S$ stochastic process if for any $n \geq 1$ and distinct indices $t_1, \ldots, t_n \in T$, the random variables $X(t_1), \ldots, X(t_n)$ are jointly $S\alpha S$ with the same characteristic exponent α.

Without loss of generality, we shall assume that all $S\alpha S$ distributions are centered at the origin so that $a = 0$. In this case, a $S\alpha S$ distribution is determined by two parameters, $0 < \alpha \leq 2$ and $\gamma > 0$, through its characteristic function

$$\varphi_{\alpha,\gamma}(t) = \exp(-\gamma |t|^\alpha). \tag{3.3}$$

Its density and distribution functions are denoted by $f_{\alpha,\gamma}(x)$ and $F_{\alpha,\gamma}(x)$, respectively. From (2.9), one can easily show that

$$f_{\alpha,\gamma}(x) = \begin{cases} \dfrac{1}{\pi \gamma^{1/\alpha}} \displaystyle\sum_{k=1}^{\infty} \dfrac{(-1)^{k-1}}{k!} \Gamma(\alpha k + 1) \sin\left(\dfrac{k\alpha\pi}{2}\right) \left(\dfrac{|x|}{\gamma^{1/\alpha}}\right)^{-\alpha k - 1} & 0 < \alpha < 1 \\[6pt] \dfrac{\gamma}{\pi(x^2 + \gamma^2)} & \alpha = 1 \\[6pt] \dfrac{1}{\pi \alpha \gamma^{1/\alpha}} \displaystyle\sum_{k=0}^{\infty} \dfrac{(-1)^k}{(2k)!} \Gamma\left(\dfrac{2k+1}{\alpha}\right) \left(\dfrac{x}{\gamma^{1/\alpha}}\right)^{2k} & 1 < \alpha < 2 \\[6pt] \dfrac{1}{2\sqrt{\gamma\pi}} \exp(-x^2/4\gamma) & \alpha = 2. \end{cases} \tag{3.4}$$

A $S\alpha S$ distribution is *standard* if γ is equal to 1.

3.2 FRACTIONAL LOWER-ORDER MOMENTS

Although the second-order moment of a $S\alpha S$ random variable with $0 < \alpha < 2$ does not exist, all moments of order less than α do exist and are called the *fractional lower-order moments* or FLOMs. The FLOMs of a $S\alpha S$ random variable can easily be found from its dispersion and characteristic exponent as follows:

Proposition 1 *Let X be a $S\alpha S$ random variable with zero location parameter and dispersion γ. Then*

$$\mathbf{E}(|X|^p) = C(p, \alpha) \gamma^{\frac{p}{\alpha}} \quad 0 < p < \alpha, \tag{3.5}$$

where

$$C(p, \alpha) = \frac{2^{p+1} \Gamma\left(\dfrac{p+1}{2}\right) \Gamma(-p/\alpha)}{\alpha \sqrt{\pi} \Gamma(-p/2)} \tag{3.6}$$

depends only on α and p, not on X. $\Gamma(\cdot)$ is the gamma function defined in (2.8).

This important result was first proved by Zolotarev using the Mellin-Stieljes transform [Zolotarev, 1957]. Cambanis and Miller [1981] rediscovered it by using a property of characteristic functions derived in [Wolfe, 1973]. A similar result is also true for complex stable random variables; see Masry and Cambanis [1984]

3.2 FRACTIONAL LOWER-ORDER MOMENTS

for a proof using the representation theory of stable processes. In the following, we give an elementary proof of the proposition based on basic properties of the gamma function.

Proof: Let X be a $S\alpha S$ random variable with zero location parameter and dispersion γ. We start with the elementary integral

$$\int_0^\infty \frac{1 - \cos at}{t^{p+1}} dt = |a|^p \frac{\Gamma(1-p)\cos\frac{\pi}{2}p}{p}$$

for any real constants a and $0 \leq p < 2$. Replacing the constant a by X and taking expectation of both sides, one has

$$E|X|^p = \frac{p}{\Gamma(1-p)\cos\frac{\pi}{2}p} \int_0^\infty E\left(\frac{1 - \cos Xt}{t^{p+1}}\right) dt.$$

Note that since X is symmetric and has characteristic function $\varphi(t) = \exp(-\gamma|t|^\alpha)$,

$$E(\cos Xt) = E(e^{jtX}) = \exp(-\gamma|t|^\alpha).$$

So,

$$E|X|^p = \frac{p}{\Gamma(1-p)\cos\frac{\pi}{2}p} \int_0^\infty \frac{1 - \exp(-\gamma|t|^\alpha)}{t^{p+1}} dt.$$

Note that the integral on the right-hand side of the above equation is finite only for $0 \leq p < \alpha$. Now fix $0 \leq p < \alpha$. Using integration by parts and (2.8), one can easily show that

$$\int_0^\infty \frac{1 - \exp(-\gamma|t|^\alpha)}{t^{p+1}} dt = \frac{1}{p} \gamma^{\frac{p}{\alpha}} \Gamma\left(1 - \frac{p}{\alpha}\right).$$

Hence

$$E|X|^p = \frac{1}{\cos\frac{\pi}{2}p} \frac{\Gamma(1-p/\alpha)}{\Gamma(1-p)} \gamma^{\frac{p}{\alpha}}.$$

Using the following properties of the gamma function [Ahlfors, 1979]:

$$\Gamma(z)\Gamma(1-z) = \frac{\pi}{\sin \pi z},$$

$$\Gamma(z+1) = z\Gamma(z),$$

and

$$\sqrt{\pi}\Gamma(2z) = 2^{2z-1}\Gamma(z)\Gamma\left(z + \frac{1}{2}\right),$$

one can easily show that

$$\frac{1}{\cos\frac{\pi}{2}p}\frac{\Gamma(1-p/\alpha)}{\Gamma(1-p)} = \frac{2^{p+1}}{\alpha\sqrt{\pi}}\Gamma\left(\frac{p+1}{2}\right)\frac{\Gamma(-p/\alpha)}{\Gamma(-p/2)}.$$

Hence the proposition is proved. □

3.3 NEGATIVE-ORDER MOMENTS

Recent investigation by Ma and Nikias [1995] shows that a $S\alpha S$ random variable also has finite negative-order moments.

Proposition 2 *Let X be a $S\alpha S$ random variable with zero location parameter and dispersion γ. Then the unified formula for its moments is*

$$E(|X|^p) = C(p,\alpha)\gamma^{\frac{p}{\alpha}} \quad -1 < p < \alpha \tag{3.7}$$

where $C(p,\alpha)$ has the same form as (3.6).

Proof: The p.d.f. of X was given in (2.5). Define Y such that $Y = |X|$, the p.d.f. for Y is

$$f_Y(y) = 2f_X(y) = \frac{2}{\pi}\int_0^\infty \cos(\omega y)\exp(-\gamma\omega^\alpha)d\omega, 0 \leq y < \infty. \tag{3.8}$$

From the definition of moments,

$$E(|X|^p) = E(Y^p) = \frac{2}{\pi}\int_0^\infty \left[\int_0^\infty y^p \cos(\omega y)dy\right]\exp(-\gamma\omega^\alpha)d\omega. \tag{3.9}$$

Notice the following integration identities:

$$\int_0^\infty x^p \cos(ax)dx = -\frac{1}{a^{p+1}}\Gamma(1+p)\sin\left(\frac{\pi p}{2}\right), \text{ for } a > 0, -1 < p < 0, \tag{3.10}$$

$$\int_0^\infty x^{\nu-1}\exp(-\mu x^\alpha)dx = \frac{1}{|\alpha|}\mu^{-\nu/\alpha}\Gamma(\nu/\alpha), \text{ for Re } \mu > 0, \text{Re } \nu > 0. \tag{3.11}$$

Then,

$$\int_0^\infty y^p \cos(\omega y)\,dy = -\frac{\Gamma(p+1)}{\omega^{p+1}} \sin\left(\frac{\pi p}{2}\right), \text{ for } \omega > 0, -1 < p < 0, \quad (3.12)$$

$$\int_0^\infty \omega^{-p-1} \exp(-\gamma \omega^\alpha)\,d\omega = \frac{1}{\alpha} \gamma^{p/\alpha} \Gamma(-p/\alpha), \text{ for } \gamma > 0, p < 0. \quad (3.13)$$

Therefore,

$$E(|X|^p) = -\frac{2}{\alpha \pi} \gamma^{p/\alpha} \Gamma(p+1) \Gamma(-p/\alpha) \sin(p\pi/2)$$

$$= \frac{2^{p+1} \Gamma(\frac{p+1}{2}) \Gamma(-p/\alpha)}{\alpha \sqrt{\pi} \Gamma(-p/2)} \gamma^{p/\alpha}, \text{ for } -1 < p < 0. \quad (3.14)$$

When $p = 0$, (3.7) is obvious. Therefore,

$$E(|X|^p) = \frac{2^{p+1} \Gamma(\frac{p+1}{2}) \Gamma(-p/\alpha)}{\alpha \sqrt{\pi} \Gamma(-p/2)} \gamma^{p/\alpha}, \text{ for } -1 < p < \alpha. \quad (3.15)$$

□

When X is an n dimensional spherically symmetric $S\alpha S$ random variable, a similar expression is

$$E(|X|^p) = 2^p \frac{\Gamma(\frac{p+n}{2}) \Gamma(1 - \frac{p}{\alpha})}{\Gamma(1 - \frac{p}{2}) \Gamma(\frac{n}{2})} \gamma^{p/\alpha}, \text{ for } -n < p < \alpha. \quad (3.16)$$

Example 3.1 Let X be $S\alpha S$ with unit dispersion. Then the fractional lower-order moments of X are given by

$$E|X|^p = \frac{2}{\pi} \Gamma(1 - p/\alpha) \Gamma(p) \sin \frac{\pi}{2} p.$$

Figure 3.1 shows $E|X|^p$ as functions of p for different α. □

3.4 LINEAR SPACE OF STABLE RANDOM VARIABLES

Let X be a $S\alpha S$ random variable with dispersion $\gamma > 0$ and location parameter $a = 0$. The norm of X is defined as

$$\|X\|_\alpha = \begin{cases} \gamma^{\frac{1}{\alpha}} & 1 \leq \alpha \leq 2 \\ \gamma & 0 < \alpha < 1. \end{cases} \quad (3.17)$$

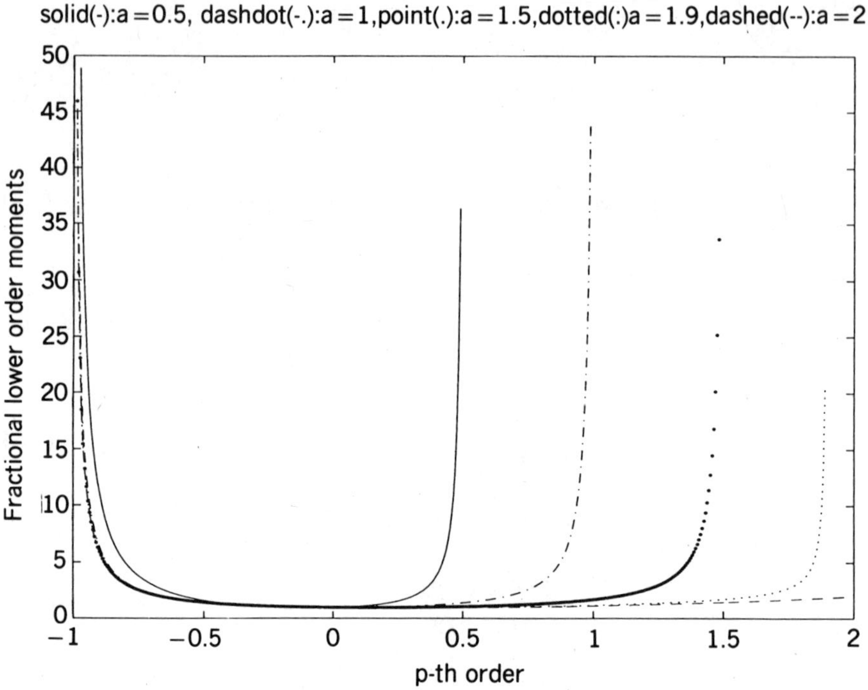

FIGURE 3.1 Fractional lower-order moments of the standardized $S\alpha S$ random variable (i.e., $\gamma = 1, a = 0$).

Thus, the norm $\|X\|_\alpha$ is basically a "scaled" version of the dispersion and determines the distribution of X via the characteristic function

$$\varphi(t) = \begin{cases} \exp\{-\|X\|_\alpha^\alpha \, |t|^\alpha\} & 1 \leq \alpha \leq 2 \\ \exp\{-\|X\|_\alpha \, |t|^\alpha\} & 0 < \alpha < 1. \end{cases}$$

If X, Y are jointly $S\alpha S$, the distance between X and Y is defined as

$$d_\alpha(X, Y) = \|X - Y\|_\alpha. \qquad (3.18)$$

Combining (3.5) and (3.17), it is easy to see that

$$d_\alpha(X, Y) = \begin{cases} [E(|X - Y|^p)/C(p, \alpha)]^{1/p} & 0 < p < \alpha, 1 \leq \alpha \leq 2 \\ [E(|X - Y|)^p/C(p, \alpha)]^{\alpha/p} & 0 < p < \alpha, 0 < \alpha < 1. \end{cases} \qquad (3.19)$$

Thus, the distance d_α between two $S\alpha S$ random variables measures the pth-order moment of the difference of these two random variables. In the case of $\alpha = 2$, this distance is half of the variance of the difference. In addition, all lower-order moments of a $S\alpha S$ random variable are equivalent, i.e., the pth and qth order moments differ by a constant factor independent of the $S\alpha S$ random variable for all $p, q < \alpha$.

It should also be mentioned that convergence in distance d_α is equivalent to convergence in probability [Cambanis and Miller, 1981]: a sequence of $S\alpha S$ random variables X_n converges to a $S\alpha S$ random variable Y in d_α if and only if it converges to Y in probability.

In the second-order moment theory, it is shown that the set of random variables with finite variance forms a Hilbert space. Similar structures exist for the set of $S\alpha S$ random variables. Let $\{X(t), t \in T\}$ be a $S\alpha S$ process. Then all finite linear combinations of elements in $\{X(t), t \in T\}$ form a linear space, denoted by $l(X(t), t \in T)$. In this space all the random variables are jointly $S\alpha S$ with the same characteristic exponent [Cambanis and Miller, 1981]. The following theorem provides a metric on the space $l(X(t), t \in T)$ [Schilder, 1970]:

Theorem 5 *For all $0 < \alpha \leq 2$, the distance defined in (3.18) is a true metric on $l(X(t), t \in T)$. In addition, for $1 \leq \alpha \leq 2$, $\|\cdot\|_\alpha$ defined in (3.17) is a norm on $l(X(t), t \in T)$.*

Let $L(X(t), t \in T)$ be the completion of $l(X(t), t \in T)$ with respect to this metric. It will be called the *linear space* of $\{X(t), t \in T\}$. It follows from the multivariate $S\alpha S$ characteristic functions that $L(X(t), t \in T)$ is a class of jointly $S\alpha S$ random variables. In particular, the random variables in the linear space of an α-sub-Gaussian process are α-sub-Gaussian [Cambanis and Miller, 1981].

As we will see later in the book, a fundamental difficulty in stable signal processing with fractional lower-order moments is that the tools of Hilbert Space Theory are no longer applicable: although the linear space of a Gaussian process is a Hilbert space, the linear space of a stable process is a Banach space for $1 \leq \alpha < 2$ and only a metric space for $0 < \alpha < 1$.[1] For both mathematical and practical reasons we will restrict α to $(1, 2]$ for the rest of the book, unless specified otherwise.

3.5 BASIC TYPES OF STABLE PROCESSES

One of the difficulties in dealing with stable processes is that there exist many types of them with mutually exclusive properties. In the following, we discuss three types of stable processes that are commonly used in practice.

3.5.1 Sub-Gaussian Processes

A stable process $\{X(t), t \in T\}$ is said to be an α-sub-Gaussian process, or briefly α-SG(R), if for all $n \geq 1$ and distinct indices $t_1, \ldots, t_n$, $(X(t_1), \ldots, X(t_n))$ has

---
[1] For a brief summary of functional analysis, see the Appendix.

characteristic function given by

$$\varphi(\mathbf{u}) = \exp\left(-\left[\frac{1}{2}\sum_{l,m=1}^{n} u_l u_m R(t_l, t_m)\right]^{\alpha/2}\right) \qquad (3.20)$$

where $R(t,s)$ is a positive definite function, $\mathbf{u} = (u_1, \ldots, u_n)^T \in R^n$, and α is restricted to $(1, 2]$. When $\alpha = 2$, $X(t)$ is of course a Gaussian process with zero mean and covariance function $R(t,s)$. Obviously, a sub-Gaussian process is stationary if and only if $R(t,s) = R(t-s) = R(s-t)$.

Sub-Gaussian processes are variance mixtures of Gaussian processes [Cambanis and Miller, 1981]. Specifically, if $X(t)$ is α-SG(R), then

$$X(t) = S^{1/2} Y(t) \qquad (3.21)$$

where S is a positive $\frac{\alpha}{2}$-stable random variable and $Y(t)$ a Gaussian process with zero mean and covariance function $R(t,s)$. In addition, S is independent of $Y(t)$. Because of this, sub-Gaussian processes are among the simplest stable processes to deal with. They share many common features with the Gaussian processes. Yet, they are also quite distinct from the Gaussian processes. For example, one of the striking properties about sub-Gaussian random variables is that they cannot be independent [Cambanis and Soltani, 1984].

Figure 3.2 shows two stationary $\frac{3}{2}$-sub-Gaussian processes with two different functions $R(\tau)$:

$$R_1(\tau) = 0.9^\tau$$

$$R_2(\tau) = 0.1^\tau.$$

Each of the two random processes is generated through (2.6), by generating a $S\frac{3}{4}S$ random variable and an AR(1) Gaussian random process with the desired correlation function $R(\tau)$. Figure 3.3 provides two more examples of α-sub Gaussian processes with the same $R(\tau) = 0.5^\tau$ and two different values of α.

3.5.2 Linear Stable Processes

Let $\{U(n), n = 0, \pm 1, \pm 2, \cdots\}$ be a family of i.i.d. $S\alpha S$ random variables. Then

$$X(n) = \sum_{i=-\infty}^{\infty} a_i U(n-i)$$

defines a stationary $S\alpha S$ random process if

$$\sum_{i=-\infty}^{\infty} |a_i|^{\alpha-\delta} < \infty$$

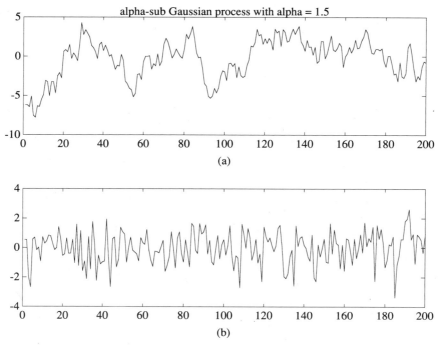

FIGURE 3.2 $\frac{3}{2}$-Sub-Gaussian process with (a) $R(\tau) = 0.9^\tau$ and (b) $R(\tau) = 0.1^\tau$.

for some $0 < \delta < \alpha$ when $0 < \alpha < 1$, or if $\sum_{i=-\infty}^{\infty} |a_i| < \infty$ when $\alpha \geq 1$ [Hosoya, 1978]. These processes are called *linear* stable processes or stable processes with moving-average representation.

Examples of linear stable processes include finite-order autoregressive (AR), moving-average (MA), and autoregressive moving-average (ARMA) processes. Specifically, let $a_1, \ldots, a_P$ be real numbers such that the zeros of the polynomial $1 - \sum_{k=1}^{P} a_k z^{-k}$ are all inside the unit circle in the complex plane. Then the following equation

$$X(n) = a_1 X(n-1) + \cdots + a_P X(n-P) + U(n), \qquad (3.22)$$

where $U(n)$'s are i.i.d. $S\alpha S$ random variables with $\alpha > 1$, has a unique stationary $S\alpha S$ solution given by

$$X(n) = \sum_{k=0}^{\infty} h_k U(n-k) \qquad (3.23)$$

where h_k is absolutely summable and

$$\frac{1}{1 - \sum_{k=1}^{P} a_k z^{-k}} = \sum_{k=0}^{\infty} h_k z^{-k} \qquad (3.24)$$

40 SYMMETRIC STABLE RANDOM PROCESSES

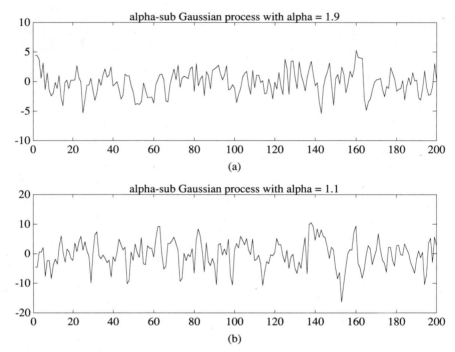

FIGURE 3.3 α-Sub-Gaussian processes with $R(\tau) = 0.5^\tau$ for (a) $\alpha = 1.9$ and (b) $\alpha = 1.1$.

on the unit circle. For a generalization of this result, see Yohai and Maronna [1977]. The process $X(n)$ thus defined is called a Pth-order AR $S\alpha S$ process. Similarly, an ARMA stable process of order (P, Q) is defined as the unique stationary solution of the following equation

$$X(n) = a_1 X(n-1) + \cdots + a_P X(n-P) + b_0 U(n) + \cdots + b_Q U(n-Q) \quad (3.25)$$

where $U(n)$'s are i.i.d. $S\alpha S$ random variables with $\alpha > 1$ and $a_1, \ldots, a_P$ are real numbers such that the zeros of the polynomial $1 - \sum_{k=1}^{P} a_k z^{-k}$ are all inside the unit circle in the complex plane. If $a_1, \ldots, a_P$ are all zero, this process is called a Qth-order MA stable process.

Figure 3.4 illustrates two AR(2) $S\alpha S$ processes with $\alpha = 2.0$ and 1.5, respectively. The two processes have the same pair of complex conjugate poles at $0.99 \pm j0.1$, which are very close to the unit circle.

Figure 3.5 shows two AR(2) $S\alpha S$ processes with $\alpha = 1.9$ and 1.1 respectively. The two processes have the same pair of real poles at 0.3 and 0.1.

Figure 3.6 shows two MA(2) $S\alpha S$ processes with $\alpha = 1.9$ and 1.1, respectively. The two processes have the same pair of complex conjugate zeros at $0.9557 \pm 0.1914j$.

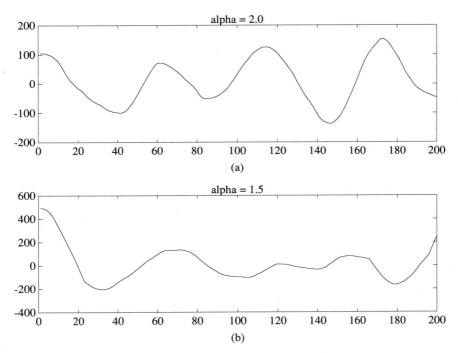

FIGURE 3.4 AR(2) $S\alpha S$ processes with a pair of complex conjugate poles close to the unit circle and with (a) $\alpha = 2.0$ and (b) $\alpha = 1.5$.

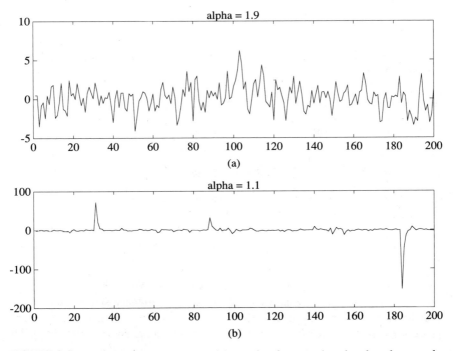

FIGURE 3.5 AR(2) $S\alpha S$ processes with a pair of unequal real poles close to the origin and with (a) $\alpha = 1.9$ and (b) $\alpha = 1.1$.

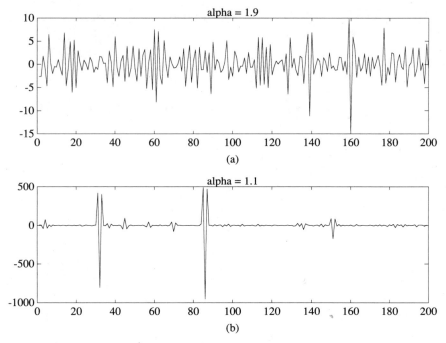

FIGURE 3.6 MA(2) $S\alpha S$ processes with a pair of complex conjugate zeros and with (a) $\alpha = 1.9$ and (b) $\alpha = 1.1$.

Figure 3.7 illustrates random processes generated by the following ARMA(2,2) model

$$X(n) = 0.195X(n-1) - 0.95X(n-2) + U(n) - 1.9114U(n-1) + 0.95U(n-2)$$

for two values of α: 1.9 and 1.1.

3.5.3 Harmonizable Stable Processes

It is well known that every wide-sense stationary second-order random sequence $\{X(n)\}$ has a spectral representation of the form

$$X(n) = \int_{-\pi}^{\pi} e^{in\omega} dZ(\omega) \qquad (3.26)$$

where $Z(\omega)$ is a second-order random process with orthogonal increments, defined on $(-\pi, \pi]$ [Papoulis, 1991]. Although not every stationary stable random process has a spectral representation, (3.26) does define an important type of stable processes called the *harmonizable* stable processes, under some appropriate inter-

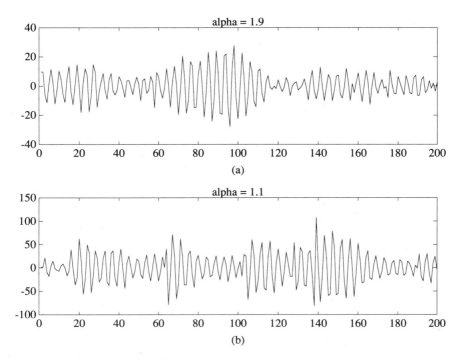

FIGURE 3.7 ARMA(2) $S\alpha S$ processes with (a) $\alpha = 1.9$ and (b) $\alpha = 1.1$.

pretation of the integral and some conditions on $Z(\omega)$. For the details, see Cambanis and Miller [1981].

It is known that the classes of linear and harmonizable Gaussian processes coincide, but in the stable case they are mutually exclusive. Namely, harmonizable stable processes have no moving-average representation and stable processes with moving-average representation are not harmonizable. In addition, sub-Gaussian processes are neither linear nor harmonizable [Cambanis and Soltani, 1984].

3.6 CONCLUSION

In this chapter we introduced the general expression for the computation of fractional lower-order moments (FLOMs) of $S\alpha S$ random variables as well as the definition and properties of the dispersion distance between two $S\alpha S$ random variables. In addition, we described three basic types of stable processes and their properties; namely, sub-Gaussian processes, linear processes based on MA, AR, and ARMA representations, and harmonizable processes.

PROBLEMS

1. Suppose that X is a random variable. For any constant a and positive constant p, prove the Chebyshev-like inequality:
$$\Pr(|X-a| \geq \epsilon) \leq \frac{E\{|X-a|^p\}}{\epsilon^p}.$$

2. The p-metric in $\mathbf{R}^n$ or $\mathbf{C}^n$ is
$$\|x\|_p \triangleq \left(\sum_{k=1}^{n} |x_k|^p\right)^{1/p}, p \geq 1.$$
   Prove that this definition satisfies the triangle inequality (Minkowski's inequality).

3. How do we define the stationarity of a $S\alpha S$ random process given that the variance is infinite?

4. (Negative-order moments!) Let X be n-dimensional ($n = 1, 2, \ldots$) spherically symmetric $S\alpha S$ random vector and Y be its length, $Y = |X| \triangleq (\sum_{i=1}^{n} x_i^2)^{1/2}$. What is the p.d.f. for Y? It has been proven that if X is an n-dimensional spherically symmetric $S\alpha S$ random vector, then its pth-order moments are finite for $0 < p < \alpha$ and
$$E\{|X|^p\} = 2^p \frac{\Gamma(1-\frac{p}{\alpha})\Gamma(\frac{n+p}{2})}{\Gamma(1-\frac{p}{2})\Gamma(\frac{n}{2})} \gamma^{p/\alpha},$$
   where γ is the dispersion. However, for a $S\alpha S$ random variable, negative-order moments also exist. Show that the unified moment expression is
$$E\{|X|^p\} = 2^p \frac{\Gamma(1-\frac{p}{\alpha})\Gamma(\frac{n+p}{2})}{\Gamma(1-\frac{p}{2})\Gamma(\frac{n}{2})} \gamma^{p/\alpha}, \underline{-n < p < \alpha}.$$

5. The tail behavior of a stable distribution was shown in Chapter 1, i.e., $\lim_{t\to\infty} t^\alpha \Pr(|X| > t) = \gamma C(\alpha)$. For every $t > 0$, find an upper bound for $t^p \Pr(|X| > t)$ where $0 < p < \alpha$.

6. Repeat Example 3.1 in the book, but change the range of p (order of the moment) to $-1 < p < \alpha$.

7. Prove the second half of Theorem 5 in Sec. 3.4 of the book, i.e., if X is a $S\alpha S$ random variable, the length of X is defined by
$$\|X\|_\alpha \triangleq \begin{cases} \gamma^{1/\alpha}, & 1 \leq \alpha \leq 2, \\ \gamma, & 0 < \alpha < 1. \end{cases}$$

Given that the above definition of $\|\cdot\|_\alpha$ is a true metric, prove that $L(X(t), t \in T)$ is a normed linear space for $1 \leq \alpha \leq 2$.

8. Suppose X_1 and X_2 are $S\alpha S$ random variables with identical density function. If they are independent, then $\|X_1 + X_2\|_\alpha = \|X_1\|_\alpha + \|X_2\|_\alpha$ for $0 < \alpha < 1$ and $(\|X_1 + X_2\|_\alpha)^\alpha = (\|X_1\|_\alpha)^\alpha + (\|X_2\|_\alpha)^\alpha$ for $1 \leq \alpha \leq 2$.

9. Show that sub-Gaussian processes are variance mixtures of Gaussian processes, i.e., if $X(t)$ is $\alpha - SG(R)$, then $X(t) \stackrel{d}{=} S^{1/2} Y(t)$, where S is a positive $\frac{\alpha}{2}$-stable random variable and $Y(t)$ is a Gaussian process and that S and $Y(t)$ are independent.

10. Can jointly sub-Gaussian random variables be independent? Justify your answer.

4

Covariation and Conditional Expectation

4.1 INTRODUCTION

The concept of covariance between two random variables plays an essential role in the second-order moment theory. The theory of signal prediction, filtering, smoothing and, in fact, most of the theory of statistical signal processing is built on covariances. Obviously, covariances do not exist on the space of $S\alpha S$ random variables, due to the lack of finite variance. Instead, a quantity called *covariation* has been proposed [Miller, 1978; Cambanis and Miller, 1981], which under certain conditions plays a role for $S\alpha S$ random variables analogous to the one played by covariance for Gaussian random variables. In the following we discuss its properties and apply it to the problem of linear regression.

4.2 COVARIATIONS AND THEIR PROPERTIES

For jointly $S\alpha S$ random variables X and Y with $1 < \alpha \leq 2$, the *covariation* of X with Y is defined by

$$[X,Y]_\alpha = \int_S xy^{\langle\alpha-1\rangle}\mu(d\mathbf{s}), \tag{4.1}$$

where S is the unit circle and $\mu(\cdot)$ is the spectral measure of the $S\alpha S$ random vector (X,Y). Here for any real number z and $a \geq 0$ we use the convention

$$z^{\langle a \rangle} = |z|^a \text{sign}(z),$$

where the function sign is defined in (2.3). In particular

$$z^{\langle 0 \rangle} = \text{sign}(z).$$

The *covariation coefficient* of X with Y is defined by

$$\lambda_{X,Y} = \frac{[X,Y]_\alpha}{[Y,Y]_\alpha}. \tag{4.2}$$

Note that the random variables X and Y play *asymmetric* roles in the above definitions.

Example 4.1 If $\alpha = 1.5$ and

$$\mu(d\phi) = d\phi$$

then

$$[X,Y]_{1.5} = \int_0^{2\pi} \cos\phi \, \sin\phi^{\langle 0.5 \rangle} d\phi = 0.$$

On the other hand, if

$$\mu(d\phi) = \left(1 - \frac{|\phi|}{2\pi}\right) d\phi$$

then

$$[X,Y]_{1.5} = \int_0^{2\pi} \cos\phi \, \sin\phi^{\langle 0.5 \rangle} \left(1 - \frac{|\phi|}{2\pi}\right) d\phi = 0.4356.$$

□

These definitions for the covariation and covariation coefficient are not very convenient in practice since they involve the use of the spectral measure $\mu(\cdot)$. Fortunately, we are able to connect the covariation and covariation coefficient with the FLOMs [Cambanis and Miller, 1981; Cambanis, Hardin, and Weron, 1988], as demonstrated in the next theorem.

Theorem 6 *Let X and Y be jointly SαS with $1 < \alpha \leq 2$. Suppose that the dispersion of Y is γ_y. Then*

$$[Y,Y]_\alpha = \|Y\|_\alpha^\alpha = \gamma_y, \tag{4.3}$$

$$\lambda_{XY} = \frac{E(XY^{\langle p-1 \rangle})}{E(|Y|^p)} \quad 1 \leq p < \alpha, \tag{4.4}$$

$$[X,Y]_\alpha = \frac{E(XY^{\langle p-1 \rangle})}{E(|Y|^p)} \gamma_y \quad 1 \leq p < \alpha. \tag{4.5}$$

We now list some of the useful properties of covariations. Their proofs can be found in Miller [1978], Cambanis and Miller [1981], and Weron [1983].

1. The covariation $[X,Y]_\alpha$ is linear in X: if X_1, X_2, Y are jointly $S\alpha S$ then

$$[aX_1 + bX_2, Y]_\alpha = a[X_1, Y]_\alpha + b[X_2, Y]_\alpha \qquad (4.6)$$

for any real constants a and b.

2. When $\alpha = 2$, i.e., when X, Y are jointly Gaussian with zero mean, the covariation of X with Y reduces to the covariance of X and Y:

$$[X,Y]_\alpha = \mathbf{E}(XY).$$

3. In general, $[X,Y]_\alpha$ is not linear with respect to the second variable Y. But it does possess the following pseudo-linearity property with respect to Y: if Y_1, Y_2 are *independent* and X, Y_1, Y_2 are jointly $S\alpha S$, then

$$[X, aY_1 + bY_2]_\alpha = a^{\langle\alpha-1\rangle}[X,Y_1]_\alpha + b^{\langle\alpha-1\rangle}[X,Y_2]_\alpha \qquad (4.7)$$

for any real constants a and b.

Example 4.2 Assume

$$\alpha = 1.5, a = 1, b = -2, [X,Y_1]_\alpha = 2, [X,Y_2]_\alpha = 3$$

Then, using the aforementioned properties of covariations, one obtains

$$[X, aY_1 + bY_2]_\alpha = 2 - 3\sqrt{2}.$$

□

4. If X, Y are independent and jointly $S\alpha S$, then

$$[X,Y]_\alpha = 0$$

while the *converse is not true in general*.

5. For any jointly $S\alpha S$ random variables X, Y, the following form of the Cauchy-Schwartz inequality holds:

$$|[X,Y]_\alpha| \leq \|X\|_\alpha \|Y\|_\alpha^{\langle\alpha-1\rangle}.$$

In particular, if X, Y have unit dispersion, one has

$$|[X,Y]_\alpha| \leq 1.$$

4.2 COVARIATIONS AND THEIR PROPERTIES

The covariation of two $S\alpha S$ random variables X and Y is, in general, difficult to calculate analytically. An important exception is when X, Y are both linear combinations of independent $S\alpha S$ random variables. Using the basic properties of covariation, it is easy to show the following proposition.

Proposition 3 *Let U_i's be independent $S\alpha S$ random variables with dispersions γ_i, $i = 1, \ldots n$. For any numbers $a_1, \ldots, a_n, b_1, \ldots, b_n$, where all b_i's are not zero, form*

$$X = a_1 U_1 + \cdots + a_n U_n,$$
$$Y = b_1 U_1 + \cdots + b_n U_n.$$

Then

$$[X, X]_\alpha = \gamma_1 |a_1|^\alpha + \cdots + \gamma_n |a_n|^\alpha,$$
$$[Y, Y]_\alpha = \gamma_1 |b_1|^\alpha + \cdots + \gamma_n |b_n|^\alpha,$$
$$[X, Y]_\alpha = \gamma_1 a_1 b_1^{\langle \alpha - 1 \rangle} + \cdots + \gamma_n a_n b_n^{\langle \alpha - 1 \rangle}, \tag{4.8}$$
$$\lambda_{XY} = \frac{\gamma_1 a_1 b_1^{\langle \alpha - 1 \rangle} + \cdots + \gamma_n a_n b_n^{\langle \alpha - 1 \rangle}}{\gamma_1 |b_1|^\alpha + \cdots + \gamma_n |b_n|^\alpha}.$$

Example 4.3 Let U be $S\alpha S$ with unit dispersion. Suppose

$$X = a_1 U$$
$$Y = U.$$

Obviously, X, Y are not independent and

$$[X, Y]_\alpha = a_1$$
$$\lambda_{XY} = a_1.$$

Thus, unlike correlation coefficients, covariation coefficients are not normalized to the interval $[-1, 1]$ and may be unbounded. □

Example 4.4 Again, let U be $S\alpha S$ with unit dispersion. Here suppose

$$X = U$$
$$Y = b_1 U.$$

Now

$$[X, Y]_\alpha = b_1^{\langle \alpha - 1 \rangle}$$
$$\lambda_{XY} = 1/b_1.$$

50 COVARIATION AND CONDITIONAL EXPECTATION

Note that λ_{XY} is independent of α and becomes unbounded as b_1 goes to zero. Figure 4.1 shows the covariation of X, Y as a function of b_1 and α. Figure 4.2 shows the covariation coefficient as a function of b_1. □

Example 4.5 Let U_1, U_2, U_3 be i.i.d. and $S\alpha S$ with $\alpha = 1.5$ and unit dispersion. Suppose

$$X = U_1 + U_2 - 2U_3,$$
$$Y = 3U_1 - 2U_2.$$

Then, their covariation coefficient is given by

$$\lambda_{XY} = \frac{\sqrt{3} - \sqrt{2}}{\sqrt{27} + \sqrt{8}}.$$

□

4.3 CONDITIONAL EXPECTATION AND LINEAR REGRESSION

As an application of the concept of covariation, let us look at the problem of linear regression with $S\alpha S$ random variables. Let $X_0, X_1, \ldots, X_n$ be jointly $S\alpha S$ random variables with $1 < \alpha \leq 2$ and spectral measure $\mu(\cdot)$. The regression of X_0 in terms of $X_1, \ldots, X_n$ is the conditional expectation $\mathbf{E}(X_0 \mid X_1, \ldots, X_n)$. It is well known that in the case where $X_0, X_1, \ldots, X_n$ are jointly Gaussian, $\mathbf{E}(X_0 \mid X_1, \ldots, X_n)$ is a linear combination of $X_1, \ldots, X_n$, thus a Gaussian random variable itself. In the $S\alpha S$ case, the regression estimate $\mathbf{E}(X_0 \mid X_1, \ldots, X_n)$ is not linear in general.

The following theorem states a necessary and sufficient condition for the regression estimate to be linear [Miller, 1978]:

Theorem 7 *If $X_0, X_1, \ldots, X_n$ are jointly $S\alpha S$ random variables with $1 < \alpha \leq 2$ and spectral measure $\mu(\cdot)$ on the unit sphere S in R^{n+1}, then*

$$\mathbf{E}(X_0 \mid X_1, \ldots, X_n) = a_1 X_1 + \cdots + a_n X_n$$

if and only if for all $r_1, \ldots, r_n$,

$$\int_S (x_0 - a_1 x_1 - \cdots - a_n x_n)(r_1 x_1 + \cdots + r_n x_n)^{\langle \alpha - 1 \rangle} \mu(d\mathbf{x}) = 0. \quad (4.9)$$

If the regression is linear, then the coefficients $a_1, \ldots, a_n$ are uniquely determined by $\mu(\cdot)$ if and only if $X_1, \ldots, X_n$ are linearly independent elements in the space of integrable random variables. For each choice of $r_1, \ldots, r_n$ the condition of the

4.3 CONDITIONAL EXPECTATION AND LINEAR REGRESSION

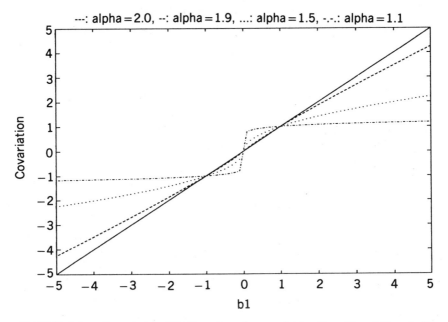

FIGURE 4.1 Covariation $[X, Y]_\alpha$ of random variables $X = U$ and $Y = b_1 U$.

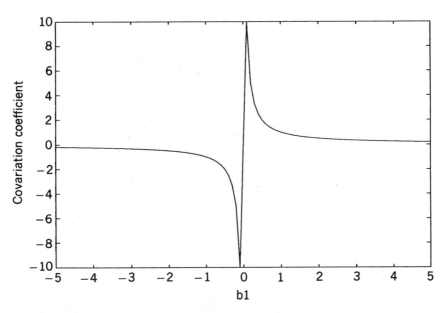

FIGURE 4.2 Covariation $[X, Y]_\alpha$ of random variables $X = U$ and $Y = b_1 U$.

theorem provides a linear equation involving the a_j's, but it is not known in general what choices of $r_1, \ldots, r_n$ will provide n linearly independent equations that can be solved for a_j's. The case $n = 2$, however, is easily solved by the following corollary [Miller, 1978]:

Corollary 1 *If X_0, X_1, X_2 are jointly SαS and*

$$E(X_0 \mid X_1, X_2) = a_1 X_1 + a_2 X_2,$$

then a_1, a_2 satisfy

$$\begin{aligned} a_1 [X_1, X_1]_\alpha + a_2 [X_2, X_1]_\alpha &= [X_0, X_1]_\alpha \\ a_1 [X_1, X_2]_\alpha + a_2 [X_2, X_2]_\alpha &= [X_0, X_2]_\alpha. \end{aligned} \quad (4.10)$$

Moreover, the above equations uniquely determine a_1, a_2 iff neither X_1 nor X_2 is a multiple of the other.

Example 4.6 Assume that X_1, X_2, and X_3 are jointly SαS and that X_3 is independent of X_1, X_2. Let

$$X_0 = a_1 X_1 + a_2 X_2 + a_3 X_3.$$

Then X_0, X_1 and X_3 are jointly SαS and

$$E(X_0 \mid X_1, X_2) = a_1 X_1 + a_2 X_2.$$

It is easy to see that a_1, a_2 satisfy (4.10). □

There are few known cases where the regression is indeed linear. The next theorem provides one example [Miller, 1978]:

Theorem 8 *If $X_0, X_1, \ldots, X_n$ are jointly SαS random variables and if $X_1, \ldots, X_n$ are independent and nondegenerate, then*

$$E(X_0 \mid X_1, \ldots, X_n) = \lambda_{01} X_1 + \cdots + \lambda_{0n} X_n$$

where λ_{0i} is the covariation coefficient of X_0 with X_i, $i = 1, \ldots, n$.

In particular,

Corollary 2 *For any two jointly SαS random variables X, Y,*

$$E(X \mid Y) = \lambda_{XY} Y. \quad (4.11)$$

Example 4.7 Assume that $Y, X_1, \ldots, X_n$ are zero-median independent $S\alpha S$ random variables, and for real numbers $a_1, \ldots, a_n$ define

$$X_0 = Y + a_1 X_1 + \cdots + a_n X_n.$$

Then, using the basic properties of expectation, one has

$$\mathbf{E}(X_0 \mid X_1, \ldots, X_n) = a_1 X_1 + \cdots + a_n X_n.$$

From Theorem 8 it follows that

$$\lambda_{0i} = a_i.$$

The same result can be obtained using the basic properties of covariations. □

Regressions of sub-Gaussian random variables are also linear. In fact, something stronger is true. A $S\alpha S$ random process $\{X(t), t \in T\}$ is said to have the *linear regression property* if $\mathbf{E}(X_0 | X_1, \ldots, X_n)$ is a linear function of $X_1, \ldots, X_n$, whenever the X_i's are elements of the linear span of $\{X(t), t \in T\}$. Then, the following remarkable result holds.

Theorem 9 *A $S\alpha S$ process has the linear regression property if and only if it is sub-Gaussian.*

The proof of this theorem is given in Hardin [1982].

4.4 COMPLEX SYMMETRIC STABLE VARIABLES

In this section, we generalize the concept of covariation to the case of complex symmetric stable random variables. Our presentation is based on the paper by Cambanis [1983].

A complex random variable $X = X_1 + jX_2$ is $S\alpha S$ if X_1, X_2 are jointly $S\alpha S$. Several complex $S\alpha S$ random variables are jointly $S\alpha S$ if their real and imaginary parts are jointly $S\alpha S$. Again, we assume that all real and complex $S\alpha S$ random variables have zero means and α is restricted to $(1, 2]$.

When $X = X_1 + jX_2$ and $Y = Y_1 + jY_2$ are jointly $S\alpha S$, the covariation of X and Y is defined by

$$[X, Y]_\alpha = \int_{S_4} (x_1 + jx_2)(y_1 + jy_2)^{\langle \alpha - 1 \rangle} d\mu_{X_1, X_2, Y_1, Y_2}(x_1, x_2, y_1, y_2) \quad (4.12)$$

where S_4 is the unit sphere in R^4 and $\mu_{X_1, X_2, Y_1, Y_2}(x_1, x_2, y_1, y_2)$ is the spectral measure of the $S\alpha S$ random vector (X_1, X_2, Y_1, Y_2). Here, for a complex number z and $\beta \rangle 0$

we use the following convention

$$z^{\langle\beta\rangle} = |z|^{\beta-1}z^* \qquad (4.13)$$

where z^* is the complex conjugate of z. The covariation coefficient of X with Y is defined by

$$\lambda_{XY} = \frac{[X,Y]_\alpha}{[Y,Y]_\alpha}. \qquad (4.14)$$

Some of the properties of real $S\alpha S$ random variables may now be extended to the complex case as follows.

Theorem 10 *Let X, Y be jointly $S\alpha S$ with zero means and $1 < \alpha \leq 2$. Then, the following properties hold [Cambanis, 1983; Cambanis and Miamee, 1989]:*

(1) $E(|X|^p) = \frac{p 2^p \Gamma(p/2)\Gamma(-p/\alpha)}{\alpha \Gamma(-p/2)}[X,X]_\alpha^{\frac{p}{\alpha}}, \quad 0 < p < \alpha$

(2) $\lambda_{XY} = \frac{E(XY^{\langle p-1\rangle})}{E(|Y|^p)}, \quad 1 < p < \alpha$

(3) $[X_1 + X_2, Y]_\alpha = [X_1, Y]_\alpha + [X_2, Y]_\alpha$

(4) $[aX, bY]_\alpha = ab^{\langle\alpha-1\rangle}[X,Y]_\alpha$

(5) $[X,Y]_\alpha = 0$ if X and Y are independent

The proof of this theorem is straightforward.

Example 4.8 Let X_1, X_2, Y_1, Y_2 be zero-mean jointly $S\alpha S$ random variables whose dispersions are $\gamma_{x1}, \gamma_{x2}, \gamma_{y1}, \gamma_{y2}$ respectively. In addition, assume that Y_1, Y_2 are independent. Let $Z = X_1 + jX_2$ and $W = Y_1 + jY_2$. Then

$$[W, W] = \gamma_{y1} + \gamma_{y2}$$

and

$$[Z, W] = [X_1, Y_1] + [X_2, Y_2] + j([X_2, Y_1] - [X_1, Y_2]). \qquad \square$$

4.5 CONCLUSION

In this chapter we discussed the concept of covariation and statistical conditional expectation of $S\alpha S$ random variables. In particular, we introduced the definition and properties of covariations and covariation coefficients along with the properties of regression estimates and conditions of linearity. Finally, the definition and key properties of complex $S\alpha S$ random variables were discussed.

PROBLEMS

1. To estimate the pth order moment $E\{|X|^p\}$, $-1 < p < \alpha$, of a $S\alpha S$ random variable X, we can use $\frac{\sum_{i=1}^{N}|x_i|^p}{N}$ as an estimator, where N is the number of i.i.d. observations. What are the mean and variance of this estimator? How are they related to p? Show the performance by Monte-Carlo simulations with $\alpha = 1.6$ and $p = -0.8, -0.2, 0.5, 1$.

2. The covariation coefficient is defined as

$$\lambda_{XY} = \frac{E(XY^{\langle p-1\rangle})}{E(|Y|^p)}, \quad \forall p, 0 \leq p < \alpha.$$

   How does the value of p affect λ_{XY} in terms of the variance of the FLOM estimator?

3. Let U_1, U_2 be i.i.d. standard $S\alpha S$ ($\alpha = 1.5$) random variables. X, Y are linear combinations of U_1, U_2:

$$X = a_1 U_1 + a_2 U_2,$$
$$Y = b_1 U_1 + b_2 U_2.$$

   Let $(a_1, a_2, b_1, b_2) = (-0.75, 0.25, 0.18, 0.78)$. What is the true value of λ_{XY}? Estimate λ_{XY} by FLOMs with $p = 0.3, 0.5, 0.7, 1.0, 1.3$. What about $\alpha = 1$ (Cauchy)?

4. This problem is about conditional mean and conditional variance.

   (a) From the linear regression: $E\{X|Y\} = \lambda_{XY} Y$, someone suggested that by taking expectation with respect to Y of both sides, one has $\lambda_{XY} = \frac{E\{X\}}{E\{Y\}}$ (assuming $E\{X\}$ and $E\{Y\}$ exist), so the covariation coefficient can be estimated accordingly. Show why this method will not work. (*Hint*: What assumption(s) does one need to get the linear regression equation: $E\{X|Y\} = \lambda_{XY} Y$?)

   (b) First define $\mathrm{Var}(X|Y)$, then prove:

$$\mathrm{Var}(X) = E\{\mathrm{Var}(X|Y)\} + \mathrm{Var}(E\{X|Y\}).$$

5. Assume X_i are i.i.d. Cauchy random variables. $X_{(n)}$ is the largest among $X_1, X_2, \ldots X_n$. Show:

$$\Pr\left(\frac{X_{(n)}}{n} \leq x\right) \to e^{-\frac{\gamma}{\pi x}},$$

   where γ is the dispersion.

6. Let X and Y be independent $S\alpha S$ ($\alpha \geq 1$) random variables with zero-location parameter, same α, and dispersions γ_x and γ_y. Random variables W, Z are

obtained by the linear transform on X, Y:

$$\begin{pmatrix} W \\ Z \end{pmatrix} = A \begin{pmatrix} X \\ Y \end{pmatrix},$$

where

$$A = \begin{pmatrix} a_{11} a_{12} \\ a_{21} a_{22} \end{pmatrix}.$$

What is the covariation matrix of X and Y? What is the covariation matrix of W and Z? How are they related to each other? (consider $\alpha = 2$ and $\alpha < 2$).

7. Based on the analogy between covariation and covariance, can we define "wide-sense" stationary stable processes?

8. X and Y are $S\alpha S$ random variables. Let $Z = aX + bY$. If X and Y are independent, then the dispersion of Z is $\gamma_z = |a|^\alpha \gamma_x + |b|^\alpha \gamma_y$. However, if X and Y are jointly $S\alpha S$ random variables for general, someone derives the following relationship between covariation and dispersion:

$$\gamma_z = [Z, Z]_\alpha = a[X, aX + bY]_\alpha + b[Y, aX + bY]_\alpha$$
$$= aa^{\langle \alpha - 1 \rangle}[X, X]_\alpha + ab^{\langle \alpha - 1 \rangle}[X, Y]_\alpha + ba^{\langle \alpha - 1 \rangle}[Y, X]_\alpha + bb^{\langle \alpha - 1 \rangle}[Y, Y]_\alpha.$$

Nice as it looks, it is wrong. Why?

5

Parameter Estimates for Symmetric Stable Distributions

5.1 INTRODUCTION

Recall that a symmetric stable distribution is determined by three parameters: the characteristic exponent α with $0 < \alpha \leq 2$, the dispersion γ with $\gamma > 0$, and the location parameter a with $-\infty < a < \infty$. A practical problem is to estimate these three parameters from the realizations of a symmetric stable random variable. For convenience, we shall replace γ by a new parameter c, defined by

$$c = \gamma^{1/\alpha} \qquad (5.1)$$

when we discuss parameter estimation methods.

For $\alpha > 1$, the sample mean will provide a consistent estimate for the location parameter a. But the problem of estimating the parameters of a stable distribution is, in general, severely hampered by the lack of known closed-form density functions for all but a few members of the stable family. Most of the conventional methods in mathematical statistics cannot be used in this case, since these methods depend on an explicit form for the density. However, there are some numerical methods that have been suggested and are discussed in this section.

5.2 METHOD OF MAXIMUM LIKELIHOOD

Approximate maximum likelihood estimates of α and γ (assuming $a = 0$) were obtained by DuMouchel [1971]. A multinomial approximation to the likelihood function is used by this approach. The estimates have the usual desirable properties of maximum likelihood estimates. However, the computational effort involved seems considerable.

A direct method can be formulated as follows. Following Zolotarev [1986] and Brorsen and Yang [1990], the standard symmetric stable density function is given by

$$f_\alpha(x) = \frac{\alpha}{|1-\alpha|\pi} x^{1/(\alpha-1)} \int_0^{\pi/2} v(\theta) e^{-x^{\alpha/(\alpha-1)} v(\theta)} d\theta \quad \text{for } \alpha \neq 1, x > 0 \quad (5.2)$$

where

$$v(\theta) = \frac{1}{(\sin \alpha\theta)^{\alpha/(\alpha-1)}} \cos[(\alpha-1)\theta](\cos \theta)^{1/(\alpha-1)}. \quad (5.3)$$

Furthermore

$$\begin{aligned} f_1(x) &= \frac{1}{\pi(1+x^2)}, \\ f_\alpha(0) &= \tfrac{1}{\pi}\Gamma((\alpha+1)/\alpha), \\ f_2(x) &= \tfrac{1}{2\sqrt{\pi}} e^{-x^2/4}. \end{aligned} \quad (5.4)$$

Therefore, the parameters α, a, c can be estimated from the observations $x_1, x_2, \ldots, x_N$ by maximizing the log likelihood function:

$$\begin{aligned} \sum_{i=1}^N \log[f_\alpha(z_i)] &= N \log \alpha - N \log(\alpha-1)\pi + \sum_{i=1}^N (\log z_i)/(\alpha-1) \\ &+ \sum_{i=1}^N \log \int_0^{\pi/2} v(\theta) e^{-z_i^{\alpha/(\alpha-1)} v(\theta)} d\theta, \end{aligned} \quad (5.5)$$

where

$$z_i = |x_i - a|/c.$$

To avoid the discontinuity and nondifferentiability at $\alpha = 1$, α is restricted to be greater than 1. Caution must be used when evaluating the integral in (5.2) and (5.5), since the integrand is singular at $\theta = 0$.

Based on (5.5), Brorsen and Yang [1990] performed Monte Carlo simulations with fairly good results. An obvious disadvantage of this method is that it is a highly nonlinear optimization problem and no initialization and convergence analysis is available.

5.3 METHOD OF SAMPLE FRACTILES

Before introducing the method of sample fractiles, we need to define the concepts of the f fractile and order statistics. Let $F(\cdot)$ be a distribution function. Then, its

f ($0 < f < 1$) *fractile*, x_f, is defined by

$$F(x_f) = f. \tag{5.6}$$

The *order statistics* of a random sample $X_1, \ldots, X_N$ are the sample values placed in ascending order. They are denoted by $X_{(1)}, \ldots, X_{(N)}$.

The order statistics are random variables that satisfy $X_{(1)} \leq \cdots \leq X_{(N)}$. In particular,

$$X_{(1)} = \min_{1 \leq i \leq N} X_i,$$

$$X_{(2)} = \text{second smallest } X_i,$$

$$\vdots$$

$$X_{(N)} = \max_{1 \leq i \leq N} X_i.$$

Thus, sample values such as the smallest, largest, or middle observation from a random sample are examples of order statistics.

Let $X_1, \ldots, X_N$ be a random sample from an unknown distribution $F(x)$, whose order statistics are $X_{(1)}, \ldots, X_{(N)}$. Given $0 < f < 1$, a consistent estimate of the f fractile, $\hat{x}_f$, is given by $X_{(N+1)f}$. As pointed out in McCulloch [1986], to avoid spurious skewness in $\hat{x}_f$, a correction must be made. Specifically, assuming that $0 \leq i \leq N$ and $\frac{2i-1}{2N} \leq f < \frac{2i+1}{2N}$, then

$$\hat{x}_f = X_{(i)} + (X_{(i+1)} - X_{(i)}) \frac{f - q(i)}{q(i+1) - q(i)} \tag{5.7}$$

where

$$q(i) = \frac{2i - 1}{2N}.$$

If $i = 0$ or $i = N$, then $\hat{x}_f = X_{(1)}$ and $\hat{x}_f = X_{(N)}$, respectively.

The most frequently used method to estimate the parameters of the $S\alpha S$ law with $1 \leq \alpha \leq 2$ is suggested by Fama and Roll [1971], and is based on order statistics. In particular, they suggested estimating c by

$$\hat{c} = \frac{1}{1.654} [\hat{x}_{0.72} - \hat{x}_{0.28}] \tag{5.8}$$

where $\hat{x}_f$ ($f = 0.72, 0.28$) is the estimated f fractile of the $S\alpha S$ distribution. They showed that the estimate of c given by (5.8) has an asymptotic bias of less than 0.4% and is asymptotically normal with variance

$$\sigma^2(\hat{c}) \approx \frac{0.09}{N[f_\alpha(0.72)]^2}, \tag{5.9}$$

where $f_\alpha(0.72)$ is the density of the distribution of X at the 0.72th fractile of the standard stable distribution of characteristic exponent α.

The characteristic exponent α, on the other hand, can be estimated from the tail behavior of the distribution. Specifically, for some large f ($f = 0.95$, for example), we first calculate

$$\hat{z}_f = \frac{\hat{x}_f - \hat{x}_{1-f}}{2\hat{c}} = 0.827 \frac{\hat{x}_f - \hat{x}_{1-f}}{\hat{x}_{0.72} - \hat{x}_{0.28}} \tag{5.10}$$

from the sample of variables. Given that X is $S\alpha S$ with characteristic exponent α and dispersion $\gamma = c^\alpha$, $\hat{z}_f$ is an estimator of the f fractile of the standard $S\alpha S$ distribution. Thus an estimate, $\hat{\alpha}_f$, can be obtained by searching a table of standard $S\alpha S$ distribution functions, such as those in Fama and Roll [1971] and Holt and Crow [1973]. Monte Carlo simulations suggest that $\hat{\alpha}_{0.95}$ and $\hat{\alpha}_{0.97}$ are fairly robust. If the true value of α is close to 2 ($\alpha > 1.9$) then the best estimator is $\hat{\alpha}_{0.99}$, in terms of both low bias and standard deviation. A new and closed-form estimator for the characteristic exponent α of a $S\alpha S$ distribution was recently proposed in Tsihrintzis and Nikias [1994] and is based on the asymptotic theory of extreme order statistics. The advantage of the new algorithm lies in the fact that unlike maximum likelihood estimators, it is robust against the stability assumption and estimates the heaviness of the distribution tails correctly in the nonstable case.

For $1 < \alpha \leq 2$, a $S\alpha S$ distribution has finite mean. Thus, the sample mean is a consistent estimate of the location parameter a. A more robust estimate is the truncated mean. A p-percent truncated sample mean is the arithmetic mean of the middle p-percent of the ranked observations. It has been shown by Monte Carlo simulations that the truncated mean estimate is very efficient and asymptotically unbiased [Fama and Roll, 1968; DuMouchel, 1975]. It has also been found that the 50% truncated mean works well when the range of α is unknown.

Fama-Roll's method is simple but suffers from a small asymptotic bias and is not asymptotically efficient. Also, α is restricted to $1 \leq \alpha \leq 2$. McCulloch [1986], generalized Fama-Roll's method to provide consistent estimates for α and c. He also eliminated the asymptotic bias in the Fama-Roll estimators of α and c. Specifically, for the symmetric stable law, the fractile estimate

$$\hat{v}_\alpha = \frac{\hat{x}_{0.95} - \hat{x}_{0.05}}{\hat{x}_{0.75} - \hat{x}_{0.25}} \tag{5.11}$$

is independent of both c and a. Thus, a consistent estimate $\hat{\alpha}$ can be found by searching tables, such as those in McCulloch [1986], with matched value of $\hat{v}_\alpha$. For fixed α, the following quantity

$$v_c = \frac{\hat{x}_{0.75} - \hat{x}_{0.25}}{c}, \tag{5.12}$$

as a function of α, is independent of a. Since $\hat{\alpha}, \hat{x}_{0.75}, \hat{x}_{0.25}$ are all consistent estimators, the following parameter is a consistent estimator of c:

$$\hat{c} = \frac{\hat{x}_{0.75} - \hat{x}_{0.25}}{v_c(\hat{\alpha})}. \tag{5.13}$$

McCulloch's method is actually more general than what we have presented here. It provides consistent estimators for all four parameters, with $-1 \leq \beta \leq 1$ and $\alpha \geq 0.6$ while retaining the computational simplicity of Fama-Roll's method.

5.4 METHOD OF SAMPLE CHARACTERISTIC FUNCTIONS

The sample characteristic function is defined as

$$\hat{\varphi}(t) = \frac{1}{N} \sum_{k=1}^{N} \exp(jtx_k) \tag{5.14}$$

where N is the sample size, and $x_1, \ldots, x_N$ are the observations. It is a consistent estimator of the true characteristic function that uniquely determines the density function. Note that the sample characteristic function, $\{\hat{\varphi}(t), -\infty < t < \infty\}$, is a stochastic process (nonstationary) with the useful property that $0 < |\hat{\varphi}(t)| \leq 1$. So, all of the moments of $\hat{\varphi}(t)$ are finite.

Several estimation methods have been proposed based on the sample characteristic function. Among these are the method of moments of Press [1972], the method of Paulson, Holcomb, and Leitch [1975], and the regression-type method of Koutrouvelis [1980, 1981]. It has been shown through simulations that Koutrouvelis's regression-type method is better than the other two in terms of consistency, bias, and efficiency [Akgiray and Lamoureux, 1989]. In the following, we describe Koutrouvelis's estimation method.

The Koutrouvelis's regression method is based on the following relations between the characteristic function of a $S\alpha S$ distribution and its parameters:

$$\log(-\log|\varphi(t)|^2) = \log(2c^\alpha) + \alpha \log|t| \tag{5.15}$$

and

$$\frac{\text{Im}\varphi(t)}{\text{Re}\varphi(t)} = \tan at. \tag{5.16}$$

From (5.15), the parameters α and c can be estimated from the linear regression

$$y_k = \mu + \alpha w_k + \varepsilon_k, \quad k = 1, 2, \ldots, K \tag{5.17}$$

where

$$y_k = \log(-\log|\hat{\varphi}(t_k)|^2), \quad \mu = \log(2c^\alpha), \quad w_k = \log|t_k|,$$

ε_k denotes an error term that is assumed to be i.i.d. with mean zero. $t_1, \ldots, t_K$ is an appropriate set of real numbers.

The location parameter a can be estimated in a similar way, through the following linear regression:

$$z_k = au_k + \varepsilon_k, \quad l = 1, 2, \ldots, L, \tag{5.18}$$

where

$$z_k = \arctan(\mathrm{Im}(\hat{\varphi}(u_k))/\mathrm{Re}(\hat{\varphi}(u_k))),$$

and $u_1, \ldots, u_L$ is an appropriate set of real numbers. The error terms ε_k are again assumed to be i.i.d. with mean zero.

The whole procedure may be performed iteratively until some prespecified convergence criterion is satisfied. The initial estimates may be provided by Fama and Roll's method or McCulloch's method. For the details about the implementation of the regression estimators, see Koutrouvelis [1980, 1981].

The regression estimators $\hat{\alpha}, \hat{c}$, and $\hat{a}$ described above are consistent and asymptotically unbiased. According to the simulation results in Koutrouvelis [1980], the regression-type method is better than the maximum likelihood method and Fama and Roll's fractile method. This method involves minimal computational effort and is easy to implement.

5.5 TESTS FOR INFINITE VARIANCE

In some applications, it may not be necessary to know the exact value of α but it is important to know whether the stable distribution is Gaussian ($\alpha = 2$) or non-Gaussian ($\alpha < 2$). For example, to decide whether to use the minimum mean-squared error criterion or the minimum dispersion criterion in adaptive filtering (see Chapter 8), we only need to know whether the stable distribution is Gaussian or non-Gaussian. Of course, a more general and difficult problem is to test if a set of data comes from a stable distribution or a nonstable distribution. For simplicity, we shall assume that the population distribution is stable and look at procedures for distinguishing non-Gaussian stable distributions from the Gaussian distribution. Since a property that differentiates the Gaussian and non-Gaussian distributions is that non-Gaussian stable distributions do not have finite variance, such tests are also called *tests for infinite variance*.

A statistically optimal test is the generalized likelihood ratio test. For the problem of distinguishing the Gaussian distribution from non-Gaussian stable distributions, this amounts to finding the ML estimate of the characteristic exponent α from the data. If $\alpha = 2$, the population distribution is Gaussian, otherwise it is non-Gaussian stable. However, this test is computationally very intensive. Hence, other heuristic, but computationally convenient, tests are used in practice.

One approach is to graphically view the *normal probability plot* of the observations [Myers, 1989]. If the plot deviates substantially from a straight line, the population of data is not normal, and hence has infinite variance. A second method is to directly test if the population distribution has finite variance. Specifically, let $X_k, k = 1, \ldots, N$, be samples from the same stable distribution. For each $1 \leq n \leq N$, form the sample variance based on the first n observations

$$S_n^2 = \frac{1}{n} \sum_{k=1}^{n} (X_k - \bar{X}_n)^2$$

where

$$\bar{X}_n = \frac{1}{n} \sum_{k=1}^{n} X_k$$

and plot the sample variance estimate S_n^2 against n. If the population distribution $F(x)$ has a finite variance, S_n^2 should converge to a finite value. Otherwise, S_n^2 will diverge. This test is called the *converging variance test* [Granger and Orr, 1972].

Example 5.1 Four sets of random $S\alpha S$ samples are generated for $\alpha = 2.0, 1.9, 1.1, 0.5$, each with 200 samples. The running sample variances are shown in Figs. 5.1(a)–(d). As expected, except for the case $\alpha = 2$, none of the running sample variances converge. □

A third test is called the *log-tail test*, proposed by Mandelbrot [1963] and also discussed in Granger and Orr [1972]. This test examines the shape of the tails of the estimated distribution function. The idea is to plot the estimate of log Prob($X > u$) against log u, where X is the random variable whose distribution is being estimated. Recall that if the true distribution is α-stable, the plot should be a straight line with slope $-\alpha$, provided u is sufficiently large (see (2.20)). Only stable distributions will have this property. If $0 < \alpha < 2$, the distribution is non-Gaussian. This test also provides an estimate of α provided that $\alpha < 2$.

Note that all of the aforementioned three tests rely on visual inspection and thus are not truly statistical tests. For other tests of the characteristics exponent see Pereira [1990], Dumouchel [1983], and Fama and Roll [1971].

5.6 SIMULATIONS

This section introduces numerical simulations using McCulloch's fractile method and Koutrouvelis's regression method for estimating the parameters of a stable distribution. For simplicity, both methods are implemented for the $S\alpha S$ case only. The reader is referred to the original papers [Koutrouvelis, 1980, 1981; McCulloch, 1986] for the estimation of the skewness index β.

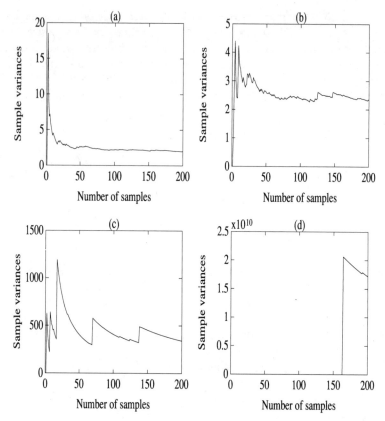

FIGURE 5.1 Running sample variances for four different values of α: (a) $\alpha = 2.0$; (b) $\alpha = 1.9$; (c) $\alpha = 1.1$; (d) $\alpha = 0.5$.

The location parameter a is estimated as the 75% truncated mean in this section. Simulations show that this estimator for a is fairly efficient for $\alpha > 0.8$. If $\alpha < 0.8$, the 25% truncated mean is more efficient.

McCulloch's fractile method for estimating α and c of a $S\alpha S$ distribution is based on the computation of four sample quantiles and simple linear interpolations of tabulated index numbers. Specifically, let the observations be $X_1, \ldots, X_N$. The first step is to calculate the fractiles $\hat{x}_{0.05}, \hat{x}_{0.25}, \hat{x}_{0.75}, \hat{x}_{0.95}$ using (5.7). The following quantity

$$\hat{v}_\alpha = \frac{\hat{x}_{0.95} - \hat{x}_{0.05}}{\hat{x}_{0.75} - \hat{x}_{0.25}}$$

is then computed. From $\hat{v}_\alpha$, an estimate of α can be found from Table 5.1 using linear interpolations.

TABLE 5.1 α versus v_α

v_α	α
2.439	2.0
2.5	1.916
2.6	1.808
2.7	1.729
2.8	1.664
3.0	1.563
3.2	1.484
3.5	1.391
4.0	1.279
5.0	1.128
6.0	1.029
8.0	0.896
10.0	0.818
15.0	0.698
25.0	0.593

TABLE 5.2 α versus v_c

α	v_c
2.00	1.908
1.90	1.914
1.80	1.921
1.70	1.927
1.60	1.933
1.50	1.939
1.40	1.946
1.30	1.955
1.20	1.965
1.10	1.980
1.00	2.000
0.90	2.040
0.80	2.098
0.70	2.189
0.60	2.337
0.50	2.588

The scale parameter c can be estimated from

$$\hat{c} = \frac{\hat{x}_{0.75} - \hat{x}_{0.25}}{\hat{v}_c(\hat{\alpha})}$$

where $\hat{v}_c(\hat{\alpha})$ is found from Table 5.2, again using linear interpolations for values not shown on the table.

The regression method of Koutrouvelis starts with an initial estimate of the parameters and proceeds iteratively until the change in the estimated values between two successive iterations is sufficiently small. The initial estimates are provided by McCulloch's fractile method. The location parameter is still estimated by the 75% truncated mean, $\hat{a}_{75}$.

Let the observations be $X_1, \ldots, X_N$ and let the current estimates be $\hat{\alpha}_i, \hat{c}_i$. To update the estimates, we first normalize the data by

$$X'_k = (X_k - \hat{a}_{75})/\hat{c}_i, \quad k = 1, \ldots, N.$$

The sample characteristic function

$$\hat{\varphi}(t) = \frac{1}{N} \sum_{k=1}^{N} \exp(jtX'_k)$$

at $t = \pi l/25, l = 1, \ldots, L$, is computed, where the optimal choice of L depends on the sample size and current estimate of α [Koutrouvelis, 1980]. For simplicity, we fix $L = 15$. The new estimates $\hat{\alpha}_{i+1}$ and $\hat{c}_{i+1}$ can be found by solving the following overdetermined system of linear equations

$$y_l = \mu + \alpha w_l, \quad l = 1, \ldots, L$$

using the least-squares, where

$$w_l = \log |t_l|, \quad l = 1, \ldots, L$$
$$y_l = \log(-\log |\hat{\varphi}(t_l)|), \quad l = 1, \ldots, L$$

and

$$\mu = \log(2c^\alpha).$$

These two methods were tested on pseudo-random $S\alpha S$ samples of length $N = 100$ and $N = 1000$ for various values of α. The parameters c and a are fixed to be 1 and 0, respectively. For each data length and each value of α, the whole procedure was repeated 50 times and the means and standard deviations of the estimates were computed. The results are shown in Table 5.3 and Table 5.4 with standard deviations in parentheses.

Both the fractile method and the regression method seem to provide accurate estimates of the characteristic exponent α and the scale parameter c. The regression method is slightly more accurate and it works for any $\alpha > 0$. The fractile method, on the other hand, applies only if $\alpha \geq 0.6$. The 75% truncated mean also provides good estimates for the location parameter when α is larger than 0.8.

5.7 ALTERNATIVE METHODS BASED ON NEGATIVE-ORDER MOMENTS

By applying the negative-order moments concept, several new methods for $S\alpha S$ parameter estimation have been proposed [Ma and Nikias, 1995]. Let X be a $S\alpha S$ random variable. It holds that

$$E|X|^p E|X|^{-p} = C(p, \alpha)C(-p, \alpha) = \frac{2\tan(p\pi/2)}{\alpha \sin(p\pi/\alpha)}, \quad 0 < p < \min(\alpha, 1), \quad (5.19)$$

where $C(p, \alpha)$ was defined in (3.6). Hence,

$$\frac{\sin(\frac{p\pi}{\alpha})}{(\frac{p\pi}{\alpha})} = \frac{2\tan(p\pi/2)}{p\pi E|X|^p E|X|^{-p}}, \quad 0 < p < \min(\alpha, 1), \quad (5.20)$$

5.7 ALTERNATIVE METHODS BASED ON NEGATIVE-ORDER MOMENTS

TABLE 5.3 Sampling Properties of the Fractile and Regression Estimates with Data Length $N = 100$

α	$\hat{a}$	Method	$\hat{\alpha}$	$\hat{c}$
2.0	−0.04399 (0.14002)	Fractile	1.85209 (0.18011)	0.98423 (0.11771)
		Regression	1.90989 (0.14431)	0.93689 (0.09950)
1.9	−0.05311 (0.14105)	Fractile	1.84852 (0.17708)	1.01767 (0.13007)
		Regression	1.81459 (0.16197)	0.94704 (0.11398)
1.7	−0.70965 (0.14832)	Fractile	1.62106 (0.19764)	1.00386 (0.11641)
		Regression	1.64972 (0.14490)	0.95635 (0.11168)
1.5	−0.22647 (0.18165)	Fractile	1.49193 (0.20005)	1.01378 (0.12571)
		Regression	1.52935 (0.17053)	0.97856 (0.10880)
1.3	−0.019183 (0.19133)	Fractile	1.25352 (0.18603)	1.02364 (0.16695)
		Regression	1.27963 (0.15199)	0.98834 (0.13411)
1.1	−0.092922 (0.19612)	Fractile	1.06466 (0.15060)	1.01619 (0.14269)
		Regression	1.10567 (0.12499)	0.97431 (0.13463)
0.9	−0.080268 (0.20133)	Fractile	0.86315 (0.11706)	0.99574 (0.18025)
		Regression	0.88200 (0.11294)	0.96556 (0.17234)
0.7	−0.13610 (0.31259)	Fractile	0.68549 (0.09548)	1.09356 (0.27332)
		Regression	0.69948 (0.09982)	0.98810 (0.20078)

TABLE 5.4 Sampling Properties of the Fractile and Regression Estimates with Data Length $N = 1000$

α	$\hat{a}$	Method	$\hat{\alpha}$	$\hat{c}$
2.0	−0.00477 (0.04648)	Fractile	1.95557 (0.06407)	0.99973 (0.04124)
		Regression	1.99060 (0.05148)	0.99354 (0.04012)
1.9	−0.00114 (0.05036)	Fractile	1.88750 (0.09589)	0.99567 (0.04726)
		Regression	1.90088 (0.06415)	0.99322 (0.04389)
1.7	−0.00021 (0.04875)	Fractile	1.70138 (0.08896)	1.00641 (0.04212)
		Regression	1.69710 (0.05933)	0.99920 (0.03605)
1.5	−0.00685 (0.05069)	Fractile	1.48979 (0.06896)	0.993286 (0.03743)
		Regression	1.48840 (0.05781)	0.98980 (0.03325)
1.3	0.00068 (0.05051)	Fractile	1.31027 (0.04848)	1.00803 (0.03427)
		Regression	1.30528 (0.04570)	1.00071 (0.02961)
1.1	−0.00369 (0.06207)	Fractile	1.10470 (0.04846)	1.00605 (0.04820)
		Regression	1.09922 (0.04358)	0.99724 (0.04529)
0.9	0.00721 (0.06785)	Fractile	0.89378 (0.04212)	1.00308 (0.06380)
		Regression	0.89603 (0.03905)	1.00066 (0.05671)
0.7	−0.00236 (0.09573)	Fractile	0.69720 (0.02806)	1.01358 (0.07300)
		Regression	0.69929 (0.03156)	1.01862 (0.06033)

5.7 ALTERNATIVE METHODS BASED ON NEGATIVE-ORDER MOMENTS

and α can then be determined in terms of the estimated moments (positive and negative order) by Eq. (5.20), which does not involve dispersion γ. Once α is obtained, γ can be solved easily by (3.5).

Another estimator based on the same idea but with a closed form is to rewrite $\mathbf{E}(|X|^p)$ as $\mathbf{E}(e^{p \log |X|})$ (note that $\log |X|$ is bounded because the p.d.f. of X, $f(x)$ in this case is bounded at $x = 0$, i.e., the probability of $x = 0$ is 0):

$$\mathbf{E}(|X|^p) = \mathbf{E}(e^{p \log |X|}) = C(p, \alpha) \gamma^{\frac{p}{\alpha}}. \tag{5.21}$$

With the negative-order moments concept, (5.21) is then continuous at $p = 0$. Define a new random variable $Y = \log |X|$, then $\mathbf{E}(e^{pY})$ is the moment-generating function of Y, and

$$\mathbf{E}(e^{pY}) = \sum_{k=0}^{\infty} \mathbf{E}(Y^k) \frac{p^k}{k!} = C(p, \alpha) \gamma^{\frac{p}{\alpha}}. \tag{5.22}$$

Therefore, moments of Y of any order must be finite and they satisfy

$$\mathbf{E}(Y^k) = \frac{d^k}{dp^k} \left(C(p, \alpha) \gamma^{\frac{p}{\alpha}} \right) \Big|_{p=0}. \tag{5.23}$$

After simplifying the above equation, one has

$$\mathbf{E}(Y) = \mathbf{C}_e \left(\frac{1}{\alpha} - 1 \right) + \frac{1}{\alpha} \log \gamma, \tag{5.24}$$

where $\mathbf{C}_e = 0.57721566\ldots$ is the Euler's constant, α is the characteristic exponent, γ is the dispersion, and

$$\mathrm{Var}(Y) = \mathbf{E}(Y - \mathbf{E}Y)^2 = \frac{\pi^2}{6} \left(\frac{1}{\alpha^2} + \frac{1}{2} \right), \tag{5.25}$$

$$\mathbf{E}(Y - \mathbf{E}Y)^3 = 2\zeta(3) \left(\frac{1}{\alpha^3} - 1 \right), \tag{5.26}$$

where $\zeta(\cdot)$ is the Riemann Zeta function, $\zeta(3) = 1.2020569\ldots$.

$$\mathbf{E}(Y - \mathbf{E}Y)^4 = \pi^4 \left(\frac{3}{20\alpha^4} + \frac{1}{12\alpha^2} + \frac{19}{240} \right). \tag{5.27}$$

The higher-order moments of Y always exist, and from the second-order moment and above, they only involve α. This property provides a simple method for pa-

TABLE 5.5 Performance of Two Methods Based on the Negative-Order Moment Concept

Estimation Method	$\hat{\alpha}$	$\hat{\gamma}$
log $|S\alpha S|$	1.4969 (0.0522)	0.9989 (0.0385)
Negative-Order Moment ($p = 0.2$)	1.5027 (0.0536)	1.0023 (0.0423)

rameter estimation. Since we can estimate the mean and variance of Y by

$$\bar{Y} = \frac{\sum_{i=1}^{N} Y_i}{N}, \tag{5.28}$$

and

$$\hat{\sigma}_Y^2 = \frac{\sum_{i=1}^{N}(Y_i - \bar{Y})^2}{N-1}, \tag{5.29}$$

where N is the number of samples and Y_i are i.i.d. observations, α and γ can then be obtained from (5.24) and (5.25). Because Y was defined as $\log|X|$ and X is a $S\alpha S$ random variable, this method is called the $\log|S\alpha S|$ method.

Example 5.2 Consider a standard $S\alpha S$ random variable with $\alpha = 1.5$, 5000 i.i.d. samples are generated for Monte-Carlo simulations. The experiment is repeated 1000 times independently. Both methods (negative-order moment method and $\log|S\alpha S|$ method) are used to estimate the characteristic exponent α and the dispersion γ. Table 5.5 shows the results. □

5.8 CONCLUSION

The problem of estimating the parameters of a stable distribution (e.g., characteristic exponent, dispersion, location parameter) from a given sample of variables is of paramount importance in practice. In this chapter, we described the following numerical methods for the estimation of the parameters of stable distribution: maximum likelihood, sample fractiles, sample characteristic functions, and estimation methods based on negative-order moments. In addition, we described tests for infinite variance and presented simulation examples that clearly show the performance of all aforementioned methods.

PROBLEMS

1. Let X be a $S\alpha S$ random variable with zero-location parameter, characteristic exponent α, and dispersion γ. When both α and γ are unknown, the Fisher's

Information matrix does not have a closed form. Even when α is known, since X does not have a closed form p.d.f., Fisher's Information for γ is still quite complicated, except for $\alpha = 1$ (Cauchy) and $\alpha = 2$ (Gaussian). What is Fisher's Information for γ when $\alpha = 1$ and $\alpha = 2$?

2. Calculate the Maximum Likelihood Estimator for the dispersion of a $S\alpha S$ random variable with zero-location parameter when $\alpha = 2$. Repeat for $\alpha = 1$.
3. Derive the details for all the equations in Sec. 5.7.
4. (Research topic) Consider a random variable X with p.d.f. $p(x)$, which is a mixture of Gaussian and non-Gaussian $S\alpha S$:

$$p(x) = (1 - \epsilon)g(x) + \epsilon f(x; \alpha, \gamma),$$

where g(x) is the p.d.f. of the Gaussian component and $f(x; \alpha, \gamma)$ is the p.d.f. of the non-Gaussian $S\alpha S$ component. Develop a method to estimate the degree of contamination, i.e., the value of ϵ, based on the i.i.d. samples of X. (*Remark*: A widely used method for parameter estimation of a mixture distribution is the EM algorithm, but notice that $S\alpha S$ distributions do not have closed-form expressions except for a few cases.)

6
Estimation of Covariations

6.1 INTRODUCTION

Covariations (or covariation coefficients) among symmetric stable random variables play the role, in certain estimation problems, of correlations (correlation coefficients) of second-order random variables. Thus, it is important to have good (unbiased, efficient) estimators for covariations and/or covariation coefficients.

In this chapter, we discuss various methods for estimating the covariation coefficient λ_{XY} of two $S\alpha S$ random variables X, Y. The reason that we focus our attention on estimating the covariation coefficient λ_{XY} instead of the covariation $[X, Y]_\alpha$ is twofold. First, since $[Y, Y]_\alpha$ is the scale parameter in the characteristic function of Y, it can be estimated by the methods described in Chapter 5. So if we know the covariation coefficient λ_{XY}, we can easily compute $[X, Y]_\alpha$ by multiplying λ_{XY} and $\|Y\|_\alpha^\alpha$. Second, as we will see soon, in most cases we need only to know the covariation coefficient. Knowledge of covariation itself is unnecessary.

6.2 FRACTIONAL LOWER-ORDER MOMENT ESTIMATOR

By Theorem 6 in Chapter 4, the covariation coefficient of jointly $S\alpha S$ random variables X and Y with $\alpha > 1$ is given by

$$\lambda_{XY} = \frac{E(XY^{\langle p-1 \rangle})}{E(|Y|^p)}$$

for any $1 \leq p < \alpha$. This immediately suggests a method for estimating the covariation coefficient λ_{XY}, which will be called a fractional lower-order moment (FLOM) estimator. Specifically, for the independent observations $(X_1, Y_1), \ldots, (X_n, Y_n)$, we

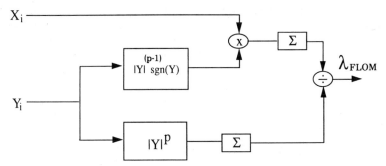

FIGURE 6.1 Implementation of the FLOM(p) estimator of covariation coefficients.

let

$$\hat{\lambda}_{FLOM(p)} = \frac{\sum_{i=1}^{N} X_i |Y_i|^{p-1} \text{sign}(Y_i)}{\sum_{i=1}^{N} |Y_i|^p} \qquad (6.1)$$

for some $1 \leq p < \alpha$. A computationally efficient choice is $p = 1$. In this case

$$\hat{\lambda}_{FLOM(p)} = \frac{\sum_{i=1}^{N} X_i \text{sign}(Y_i)}{\sum_{i=1}^{N} |Y_i|}. \qquad (6.2)$$

Figure 6.1 is a flow diagram graph that illustrates how to implement the FLOM estimator.

6.3 SCREENED RATIO ESTIMATOR

Kanter and Steiger [1974] proposed an unbiased and consistent estimator for covariation coefficients, based on the following theorem:

Theorem 11 *Let X and Y be random variables that satisfy* $\mathbf{E}(|X|) < \infty$ *and*

$$\mathbf{E}(X \mid Y) = \lambda Y \qquad (6.3)$$

for some constant λ. For $0 < c_1 < c_2 \leq \infty$ define a new random variable χ_Y from Y as follows:

$$\chi_Y = \begin{cases} 1 & \text{if } c_1 < |Y| < c_2 \\ 0 & \text{otherwise.} \end{cases}$$

Then

$$E(XY^{-1}\chi_Y)/P(c_1 < |Y| < c_2) = \lambda.$$

Thus, for independent observations $(X_1, Y_1), \ldots, (X_n, Y_n)$, the following estimate

$$\hat{\lambda}_{SCR} = \sum_{i=1}^{N}(X_i Y_i^{-1}\chi_{Y_i})/\sum_{i=1}^{N}\chi_{Y_i} \qquad (6.4)$$

is an unbiased estimate of λ. As with any unbiased estimate, the strong law of large numbers implies that the above estimate is strongly consistent. Namely, $\hat{\lambda}_{SCR}$ converges to λ almost surely as $n \to \infty$, where X_i, Y_i are independent copies of X and Y, respectively. $\hat{\lambda}_{SCR}$ in (6.4) is called the *screened ratio* (SCR) estimate of λ. The constants c_1, c_2 can be arbitrary. A common choice is to let $c_1 > 0$ and $c_2 = \infty$.

Example 6.1: Suppose we observe

X_i	0.5	1.5	2.0	3.0	4.0
Y_i	−1.1	0.5	0.8	2.1	3.0

If we choose $c_1 = 1.0, c_2 = 2.5$ then from (6.4) it follows that

$$\hat{\lambda}_{SCR} = \frac{0.2/(-1.1) + 3.0/2.1}{2} = 0.6234. \qquad \square$$

6.4 LEAST-SQUARES ESTIMATORS

A method that works well for estimating correlation coefficients of Gaussian random variables is the least-squares (LS) method. Specifically, we may estimate λ_{XY} from N observations of X_i, Y_i by minimizing the error

$$\sum_{i=1}^{N}(X_i - \lambda Y_i)^2.$$

The solution is of course the so-called least-squares estimate for the correlation coefficient of X and Y:

6.5 SAMPLING RESULTS AND SOME COMPARISONS

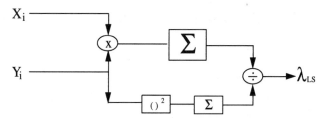

FIGURE 6.2 Implementation of least-squares estimator of covariation coefficients.

$$\hat{\lambda}_{LS} = \sum_{i=1}^{N} X_i Y_i / \sum_{i=1}^{N} Y_i^2. \tag{6.5}$$

As pointed out in Kanter and Steiger [1974], the least-squares estimate $\hat{\lambda}_{LS}$ is not consistent in the infinite variance case. We only include it for the purpose of comparison.

Figure 6.2 illustrates the flow diagram graph that shows how to implement the least-squares estimator.

6.5 SAMPLING RESULTS AND SOME COMPARISONS

In this section we compare the performances of the proposed estimators of covariation coefficients by means of Monte Carlo simulations. Two $S\alpha S$ random variables, X and Y, are generated by

$$X = a_1 U_1 + a_2 U_2,$$
$$Y = b_1 U_1 + b_2 U_2$$

where U_1, U_2 are independent, symmetrically distributed stable random variables with characteristic function $\varphi(t) = \exp(-|t|^\alpha)$. So the true covariation coefficient λ_{XY} of X with Y is

$$\lambda_{XY} = \frac{a_1 b_1^{\langle \alpha-1 \rangle} + a_2 b_2^{\langle \alpha-1 \rangle}}{|b_1|^\alpha + |b_2|^\alpha}.$$

We generated 5000 independent samples of U_1 and U_2 and the three estimators were computed as follows:

FLOM estimator:

$$\hat{\lambda}_{FLOM(p)}(N) = \frac{\sum_{i=1}^{N} X_i \operatorname{sign}(Y_i)}{\sum_{i=1}^{N} |Y_i|}. \tag{6.6}$$

Screened ratio estimator:

$$\hat{\lambda}_{SCR}(N) = \sum_{i=1}^{N}(X_i Y_i^{-1} \chi_{Y_i}) / \sum_{i=1}^{N} \chi_{Y_i} \qquad (6.7)$$

where we choose $c_1 = 0.1, c_2 = \infty$.

Least-squares estimator:

$$\hat{\lambda}_{LS}(N) = \sum_{i=1}^{N} X_i Y_i / \sum_{i=1}^{N} Y_i^2. \qquad (6.8)$$

Finally, this experiment was repeated independently 50 times, and the means and standard deviations of the three estimators were computed. The results are shown in Table 6.1 with standard deviations in parentheses.

Several remarks stand out from the simulation results illustrated in Table 6.1. The least-squares estimator works well only in the Gaussian case ($\alpha = 2$). It completely fails for $\alpha < 2$. The screened ratio (SCR) estimator performs well for $\alpha < 2$. The smaller α is, the better its performance. These observations are consistent with the theoretical result that the screened ratio estimator is unbiased and consistent. On the other hand, the FLOM(p) estimator with $p = 1$ is very robust against changes of α, although it becomes more volatile when α approaches 1. Thus the FLOM(p) estimator seems to be a very promising method for estimating covariations, especially when α is close to 2. In addition, it is very simple and inexpensive to compute.

It should be pointed out that the above remarks are valid only for independent observations. When the observations X_i, Y_i are dependent, the conclusions of these sampling results are no longer true, as we will demonstrate in the next chapter.

TABLE 6.1 Sampling Properties of Covariation Estimators

Model (a_1, a_2, b_1, b_2)	α	LS	SCR	FLOM ($p = 1$)	True λ_{XY}
	1.1	0.3340 (5.0539)	−0.4327 (0.3919)	−0.4707 (3.1266)	−0.4252
	1.3	0.3752 (5.1640)	−0.2591 (0.4991)	−0.2599 (1.0352)	−0.2602
(−0.75, 0.25, 0.18, 0.78)	1.5	0.4059 (4.5142)	−0.1112 (0.3337)	−0.1222 (0.9202)	−0.1273
	1.9	0.1069 (0.8870)	0.0654 (0.3175)	0.06104 (0.4598)	0.0599
	2.0	0.0976 (0.0102)	0.08544 (0.2790)	0.0954 (0.1252)	0.0936

6.6 CONCLUSION

Although the covariation coefficients of stable random variables play the role of correlation coefficients in many statistical problems, their behavior is quite different from that of correlation coefficients of second-order random variables. For example, although the correlation coefficients are always bounded by 1, the covariation coefficients can become unbounded. In this chapter, we introduced three different methods for the estimation of covariation coefficients from independent observations of two $S\alpha S$ random variables; namely, the fractional lower-order moment (FLOM(p)) estimator, the screened ratio (SCR) estimator, and the least-squares (LS) estimator, which is only valid for the Gaussian case ($\alpha = 2$). Finally, we presented performance comparisons by using Monte Carlo simulations.

PROBLEMS

1. (Blind Identification) For the following FIR channel:
$$Y_n = h_0 X_n + h_1 X_{n-1},$$
where h_0, h_1 are the channel impulse response coefficients and the input X_n's are i.i.d. $S\alpha S$ random variables, find the closed-form expressions for h_0 and h_1 in terms of the *output* covariations.

2. (Continuing Problem 1) Find the closed-form expressions for h_0, h_1, and h_2 in terms of the output covariations for the following FIR channel:
$$Y_n = h_0 X_n + h_1 X_{n-1} + h_2 X_{n-2}.$$

3. It is known that X and Y are two jointly $S\alpha S$ random variables. If they are independent, then their covariation $[X, Y]_\alpha = 0$. The converse is not true in general, except for the Gaussian case ($\alpha = 2$). Prove that for a linear space of $S\alpha S$ random variables with dimension higher than 1, $[X, Y]_\alpha = 0$ implies that X and Y are independent if and only if $\alpha = 2$, i.e., X and Y are zero-mean Gaussian.

4. X and Y are two jointly non-Gaussian $S\alpha S$ ($0 < \alpha < 2$) random variables with the same norm $\|X\|_\alpha = \|Y\|_\alpha$. In general, $[X, Y]_\alpha \ne [Y, X]_\alpha$ (why?). Prove that $[X, Y]_\alpha = [Y, X]_\alpha$ if and only if X and Y are jointly sub-Gaussian random variables.

5. The FLOM (fractional lower-order moment) estimator for the covariation is
$$\hat{\lambda}_{\text{FLOM}} = \frac{\sum_{i=1}^{N} X_i |Y_i|^{p-1} \text{sign}(Y_i)}{\sum_{i=1}^{N} |Y_i|^p},$$

for $1 \leq p < \alpha$. The choice of p is arbitrary. Repeat the experiment of Table 6.1 by using different p. How do the results compare?

6. The screened ratio estimator is

$$\hat{\lambda}_{\text{SCR}} = \frac{\sum_{i=1}^{N} X_i Y_i^{-1} \chi_{Y_i}}{\sum_{i=0}^{N} \chi_{Y_i}},$$

where

$$\chi_Y = \begin{cases} 1, & c_1 < |Y| < c_2, \\ 0, & \text{else}, \end{cases}$$

for $0 < c_1 < c_2 < \infty$. The choices for c_1 and c_2 are arbitrary. Repeat the experiment of Table 6.1 for different values of c_1 and c_2. How do the results compare?

7. Can we apply the FLOM estimator and the screened ratio estimator to estimate the covariance of two jointly Gaussian random variables? Perform the Monte Carlo simulations and show how the results compare with the least square estimator of the covariance.

8. When $\alpha \leq 1$, the covariation is undefined. Another measure for the relationship of two jointly $S\alpha S$ random variables is codifference, which is defined as:

$$\tau_{X,Y} = \|X\|_\alpha^\alpha + \|Y\|_\alpha^\alpha - \|X - Y\|_\alpha^\alpha,$$

where the norm of a $S\alpha S$ random variable was defined in Sec. 3.3. How do you estimate the codifference? Repeat the experiment of Table 6.1 with $\alpha \leq 1$.

7

Parametric Models of Stable Processes

7.1 INTRODUCTION

Unlike Gaussian random processes, which are completely determined by the autocorrelation sequences, stable processes are difficult to characterize. In this chapter, we consider a class of linear $S\alpha S$ processes that are generated by systems with rational transfer functions driven by i.i.d. $S\alpha S$ inputs. This class of processes includes AR, MA, and ARMA models. The output process can be completely described in terms of model parameters and the dispersion of the input. We show how to estimate the model parameters from the covariation sequence of the output. The developments here are similar to those of second-order linear processes, using second-order statistics [Marple, 1987].

7.2 PARAMETER ESTIMATION OF AR STABLE PROCESSES

Consider the process generated by the following AR system

$$X(n) = a_1 X(n-1) + \cdots + a_P X(n-P) + U(n) \qquad (7.1)$$

where $\{U(n)\}$ is a sequence of i.i.d. $S\alpha S$ random variables of characteristic exponent α and dispersion γ_u, i.e., the random variables $U(n)$ have the same characteristic function $\exp(-\gamma_u |t|^\alpha)$. As in the second-order case, we are only interested in the stationary solution of (7.1), which is unique if the AR system in (7.1) is stable, i.e., if all the poles lie inside the unit circle. In this case, the stationary solution is given by (3.23) and (3.24). Note that the stationary process $X(n)$ is $S\alpha S$ and $X(n)$ and $U(n+j)$ are independent for any $j > 0$.

We are interested in the problem of identifying the AR coefficients $a_1, \ldots, a_P$ from observations of the output $X(n)$. In the following, we review some of the approaches and compare their performances through Monte Carlo simulations.

7.2.1 Generalized Yule-Walker Equation

Taking the conditional expectation of both sides of (7.1) given $X(m)$, we obtain that for $n - P \leq m \leq n - 1$:

$$E(X(n) \mid X(m)) = a_1 E(X(n-1) \mid X(m)) + \cdots + a_P E(X(n-P) \mid X(m)) \quad (7.2)$$

where we used the fact that $E(U(n) \mid X(m)) = 0$ if $U(n)$ and $X(m)$ are independent. Since $\{X(n)\}$ is stable and stationary, we have

$$E(X(n+l) \mid X(n)) = \lambda(l) X(n)$$

where $\lambda(l)$ is the covariation coefficient of $X(n+l)$ with $X(n)$ with $\lambda(0) = 1$. Let us also define

$$\mathbf{p} = \begin{bmatrix} \lambda(1) \\ \lambda(2) \\ \vdots \\ \lambda(P) \end{bmatrix}, \quad \mathbf{a} = \begin{bmatrix} a_1 \\ a_2 \\ \vdots \\ a_P \end{bmatrix}$$

and

$$\mathbf{C} = \begin{bmatrix} \lambda(0) & \lambda(-1) & \cdots & \lambda(1-P) \\ \lambda(1) & \lambda(0) & \cdots & \lambda(2-P) \\ \vdots & \vdots & \ddots & \vdots \\ \lambda(P-1) & \lambda(P-2) & \cdots & \lambda(0) \end{bmatrix}. \quad (7.3)$$

Then the AR coefficients can be found by solving the following system of linear equations:

$$\mathbf{Ca} = \mathbf{p}. \quad (7.4)$$

Equation (7.4) is a direct generalization of the Yule-Walker equation for the case $\alpha = 2$. The covariation matrix C is Toeplitz. Thus, if it is nonsingular, we can solve (7.4) fairly efficiently. On the other hand, unlike the Gaussian case where C is symmetric and positive definite, the covariation matrix C is not symmetric and it may even be singular.

In practice, we need to estimate the covariation matrix $\mathbf{C}$ and vector $\mathbf{p}$ from the output observations. In Chapter 6, we presented three methods for estimating covariation coefficients and studied their sampling properties. We found that the FLOM and SCR estimators are consistent. In fact, the SCR estimator is strongly consistent. The LS estimator, on the other hand, is not consistent. Based on these conclusions, one would suggest the use of the FLOM and SCR estimators of the covariation coefficients in the generalized Yule-Walker equation. The only problem is that the consistency results in Chapter 6 were obtained under the assumption of independent observations. The output sequence of an AR process is far from being independent. As a consequence, although the FLOM estimator is still consistent, the SCR estimator is not. In fact, sampling studies show that the SCR estimator is so inefficient for AR processes that it is virtually useless for determining the AR coefficients. More surprising, however, is the fact that the LS estimator is consistent for AR α-stable processes and very efficient. This has been proved rigorously in Kanter and Steiger, [1974] and supported by sampling studies. Hence, if the covariation coefficients in the Generalized Yule-Walker equation are replaced by their FLOM or LS estimates, consistent estimates for the AR coefficients are obtained.

The properties of estimates of AR coefficients based on the LS estimates of covariation coefficients are further studied in the following subsection, as we show that these estimates can be obtained directly through linear least-squares regression.

7.2.2 Least-Squares Method

Another approach to estimate the AR coefficients is to treat the AR model as if it were driven by a second-order process and use the linear least-squares regression. Thus, the least-squares estimates of $a_1, \ldots, a_P$ are found by minimizing

$$\sum_{n=P}^{N} (X(n) - a_1 X(n-1) - \cdots - a_P X(n-P))^2.$$

It is easy to see that the LS estimates of $a_1, \ldots, a_P$ are the solution of the following equation (assuming $\hat{C}_{LS}$ is invertible):

$$\hat{C}_{LS} \hat{a}_{LS} = \hat{p}_{LS}, \qquad (7.5)$$

where $\hat{C}_{LS} = (\hat{\lambda}(i, j))$, $\hat{p}_{LS} = (\hat{\lambda}(i))$ and

$$\hat{\lambda}(i, j) = \sum_{n=P}^{N} X(n-j) X(n-i) \Big/ \sum_{n=P}^{N} X^2(n-i), \qquad (7.6)$$

$$\hat{\lambda}(i) = \sum_{n=P}^{N} X(n) X(n-i) \Big/ \sum_{n=P}^{N} X^2(n-i). \qquad (7.7)$$

When the data length is sufficiently large, $\hat{\lambda}(i, j) \approx \hat{\lambda}_{LS}(i-j)$ and $\hat{\lambda}(i) \approx \hat{\lambda}_{LS}(i)$, where $\hat{\lambda}_{LS}(i)$ is the LS estimate of the covariation of $X(n+i)$ and $X(n)$. Hence, asymptotically, the least-squares estimates of AR coefficients are the same as the solution of the generalized Yule-Walker equation based on the LS estimates of covariation coefficients. Since the latter have been shown to be consistent, the least-squares estimates of AR coefficient are also consistent. In fact, they are strongly consistent and the order of their convergence is given by the following theorem [Kanter and Steiger, 1974; Yohai and Maronna, 1977; Hannan and Kanter, 1977]:

Theorem 12 *Let N be the number of observations used in computing the least-squares estimates $\hat{a}_1(N), \ldots, \hat{a}_P(N)$, as shown in (7.5), (7.6), and (7.7). Then, for any $\delta > \alpha$, $\lim_{N \to \infty} N^{1/\delta}(\hat{a}_j(N) - a_j) = 0$ a.s. for $j = 1, 2, \ldots, P$.*

Hence the convergence of the least-squares estimates to the true coefficients is very rapid, on the order of $O(N^{-1/\delta})$ for any $\delta > \alpha$. The smaller α, the faster the convergence.

7.2.3 Least Absolute Deviation Estimates

It is well known that the least-squares method is closely related to the minimum mean-squared error estimation method. Hence, it is surprising that the least-squares method has such nice properties for estimating the coefficients of AR processes with infinite variance. After all, in this case, the minimum mean-squared error estimation approach does not even apply.

For $S\alpha S$ random variables, a suitable measure of dispersion is the norm defined by (3.17). Since for $p < \alpha$, the L_p norm $(\mathbf{E}(|X|^p))^{1/p}$ exists for any $S\alpha S$ random variable X and is equivalent to $\|X\|_\alpha$, we may estimate the coefficients by minimizing

$$\mathbf{E}|X(n) - a_1 X(n-1) - \cdots - a_P X(n-P)|^p, \quad 0 < p < \alpha.$$

For finite observations, we find $a_1, \ldots, a_P$ by minimizing the function

$$\sum_{n=P}^{N} |X(n) - a_1 X(n-1) - \cdots - a_P X(n-P)|^p.$$

A particularly simple case is when $p = 1$ [Blattberg and Sargent, 1971; An and Chen, 1982]. The estimates $\hat{a}_1, \ldots, \hat{a}_P$ are then obtained by minimizing

$$\sum_{n=P}^{N} |X(n) - a_1 X(n-1) - \cdots - a_P X(n-P)| \tag{7.8}$$

and are called the *least absolute deviation* (LAD) estimates. In general, the LAD estimates are unique. The consistency of LAD estimates in the case of infinite variance is established by the following theorem [Gross and Steiger, 1979; An and Chen, 1982]:

Theorem 13 *Let N be the number of observations used in computing the LAD estimates $\hat{a}_j(N)$, $j = 1, 2, \ldots, P$ of the AR coefficients. Then, for any $\delta > \alpha$, the LAD* estimates $\hat{a}_j(N)$ satisfy

$$\lim_{N\to\infty} N^{1/\delta}(\hat{a}_j(N) - a_j) = 0 \text{ in probability for } j = 1, 2, \ldots, P. \qquad (7.9)$$

Hence, the convergence rate of the LAD estimates is comparable with that of the LS estimates, although simulations show that the *LAD* estimates actually converge faster than the LS estimates [Gross and Steiger, 1979] (also see Chapter 8). But the LAD estimates are computationally more costly than the LS estimates in the present nonadaptive form.

7.2.4 Sampling Results and Performance Comparisons

To study the large sample properties of the proposed estimators, we generated 5000 samples of a stationary $S\alpha S$ output of a second-order AR system defined by

$$X(n) - a_1 X(n-1) - a_2 X(n-2) = U(n),$$

where

$$a_1 = 0.195, \quad a_2 = -0.95.$$

The estimates $\hat{a}_1, \hat{a}_2$ were computed based on the generalized Yule-Walk equation, the LS method, and the LAD method. For the generalized Yule-Walker method, we used both SCR and FLOM estimates of the covariation coefficients. This experiment was repeated independently 50 times and the means and standard deviations of the three estimates were computed. These results are shown in Table 7.1, where standard deviations are given in parentheses.

From this table, it is obvious that the LS and LAD methods are very efficient and reliable. The generalized Yule-Walker method based on FLOM estimates with $p = 1$ performs as well as the least squares despite its simplicity. This is not surprising, knowing its superior performance in estimating covariations. On the other hand, the AR parameter estimates based on the screened ratio (SCR) estimates of covariation coefficients exhibit poor performance.

To show the stability of these estimators more clearly, the corresponding AR frequency responses with the estimated parameters are shown in Figures 7.1–7.5.

7.3 PARAMETER ESTIMATION OF ARMA STABLE PROCESSES

Consider an α-stable process generated by the following ARMA equation

$$X(n) = a_1 X(n-1) + \cdots + a_P X(n-P) + b_0 U(n) + \cdots + b_Q U(n-Q), \qquad (7.10)$$

where $\{U(n)\}$ is a sequence of i.i.d. $S\alpha S$ random variables of characteristic exponent α and dispersion γ_u. Under the assumption that the system is stable, the output

TABLE 7.1 Estimation of Parameters of a Second-Order AR α-Stable Process

Coefficients	α	Least squares	LAD	Yule-Walker (FLOM)	Yule-Walker (SCR)
	1.1	0.1953 (0.0277)	0.1950 (0.0021)	0.1942 (0.1178)	0.2292 (0.8829)
	1.3	0.1946 (0.0299)	0.1949 (0.0039)	0.1949 (0.0895)	0.1882 (0.6984)
$a_1 = 0.195$	1.5	0.1944 (0.0022)	0.1945 (0.0110)	0.1929 (0.0918)	0.2054 (0.3776)
	1.9	0.1951 (0.0031)	0.1946 (0.0362)	0.1949 (0.0563)	0.2027 (0.2480)
	2.0	0.1950 (0.0347)	0.1953 (0.0437)	0.1945 (0.0392)	0.1981 (0.1847)
	1.1	−0.9486 (0.0014)	−0.9500 (0.0017)	−0.9454 (0.0049)	−0.9500 (0.0013)
	1.3	−0.9500 (0.0271)	−0.95 (0.0050)	−0.9519 (0.002)	−0.9501 (0.0057)
$a_2 = -0.95$	1.5	−0.9493 (0.0008)	−0.9498 (0.0093)	−0.9513 (0.0913)	−0.9610 (0.4647)
	1.9	−0.9500 (0.0001)	−0.95 (0.0001)	−0.9513 (0.0014)	−0.9450 (0.2600)
	2.0	−0.95 (0.00001)	−0.9503 (0.0003)	−0.9498 (0.0002)	−0.9490 (0.0010)

process $\{X(n)\}$ is a stationary $S\alpha S$ random sequence of characteristic exponent α. In addition, $X(n)$ and $U(n + j)$ are independent for all $j > 0$. The problem is to identify the AR coefficients $a_1, \ldots, a_P$ and the MA coefficients $b_0, b_1, \ldots, b_Q$ from output observations.

7.3.1 Higher-Order Yule-Walker Equation

Let us define

$$\lambda_{xx}(m) = \lambda_{X(n)X(n-m)} \tag{7.11}$$

as the covariation coefficient of $X(n)$ with $X(n - m)$ and

$$\lambda_{ux}(m) = \lambda_{U(n)X(n-m)} \tag{7.12}$$

as the covariation coefficient of $U(n)$ with $X(n - m)$. Because of the stationarity of the random processes, these covariations do not depend on the time n.

7.3 PARAMETER ESTIMATION OF ARMA STABLE PROCESSES 85

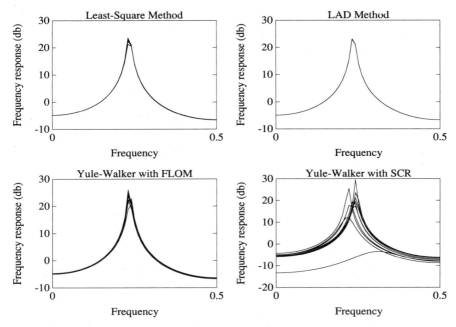

FIGURE 7.1 AR model frequency response with the estimated parameters when $\alpha = 1.1$.

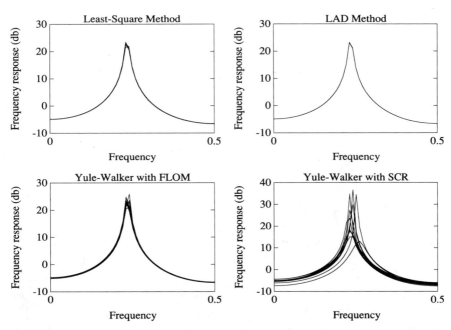

FIGURE 7.2 AR model frequency response with the estimated parameters when $\alpha = 1.3$.

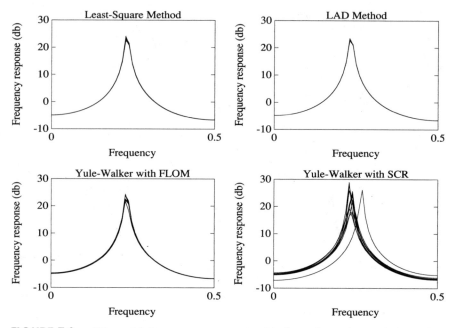

FIGURE 7.3 AR model frequency response with the estimated parameters when $\alpha = 1.5$.

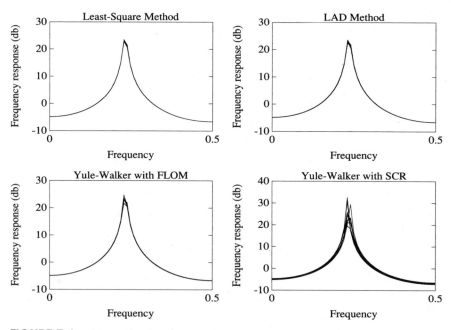

FIGURE 7.4 AR model frequency response with the estimated parameters when $\alpha = 1.9$.

7.3 PARAMETER ESTIMATION OF ARMA STABLE PROCESSES

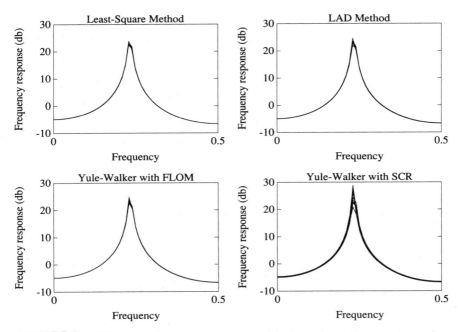

FIGURE 7.5 AR model frequency response with the estimated parameters when $\alpha = 2.0$.

By computing the covariations of both sides of (7.10) with $X(n-m)$, one obtains

$$\lambda_{xx}(m) = \sum_{k=1}^{P} a_k \lambda_{xx}(m-k) + \sum_{k=0}^{Q} b_k \lambda_{ux}(m-k). \qquad (7.13)$$

The covariation $\lambda_{ux}(m)$ between the input and output of the ARMA system can be expressed in terms of its linear impulse response $h(n)$ by using the linearity property of covariations; viz:

$$\lambda_{ux}(m) = \begin{cases} 0 & \text{for } m > 0 \\ \gamma_u & \text{for } m = 0 \\ \gamma_u [h(-m)]^{\langle \alpha-1 \rangle} & \text{for } m < 0. \end{cases} \qquad (7.14)$$

Combining (7.13) and (7.14), we obtain the relationship between the ARMA parameters and the covariation of the process $X(n)$:

$$\lambda_{xx}(m) = \begin{cases} \sum_{k=1}^{P} a_k \lambda_{xx}(m-k) + \gamma_u \sum_{k=m}^{Q} b_k [h(k-m)]^{\langle \alpha-1 \rangle} & \text{for } 0 \leq m \leq Q \\ \sum_{k=1}^{P} a_k \lambda_{xx}(m-k) & \text{for } m > Q. \end{cases}$$
$$(7.15)$$

Restricting the lag index m to $Q+1 \leq m \leq Q+P$, one obtains a set of linear equations for the autoregressive parameters $a_1, \ldots, a_P$ of an ARMA model in terms of the covariation sequence:

$$\begin{bmatrix} \lambda_{xx}(Q) & \lambda_{xx}(Q-1) & \cdots & \lambda_{xx}(Q-P+1) \\ \lambda_{xx}(Q+1) & \lambda_{xx}(Q) & \cdots & \lambda_{xx}(Q-P+2) \\ \vdots & \vdots & \ddots & \vdots \\ \lambda_{xx}(Q+P-1) & \lambda_{xx}(Q+P-2) & \cdots & \lambda_{xx}(Q) \end{bmatrix} \begin{bmatrix} a_1 \\ a_2 \\ \vdots \\ a_P \end{bmatrix} = \begin{bmatrix} \lambda_{xx}(Q+1) \\ \lambda_{xx}(Q+2) \\ \vdots \\ \lambda_{xx}(Q+P) \end{bmatrix}.$$

(7.16)

Thus, the the autoregressive parameters may be found separately from the moving average parameters as the solution of the simultaneous set of equations (7.16). This is analogous to the ARMA Yule-Walker normal equations. Note that the covariation matrix in (7.16) is again Toeplitz.

Unfortunately, the moving average parameters of an ARMA model cannot be simply found as the solution of a set of linear equations. The MA parameters are convolved with the impulse response coefficients $h(k)$, as indicated in (7.16), resulting in a nonlinear relationship to the covariation sequence. In the simplest case where $P = 0$ (i.e., for MA models), it is easy to show that the relationship between the model parameters and output covariation sequence is given by

$$\lambda_{xx}(m) = \begin{cases} \gamma_u \left(\sum_{k=0}^{Q-m} b_{k+m} b_k^{\langle \alpha-1 \rangle} \right) & \text{for } 0 \leq m \leq Q \\ \gamma_u \left(\sum_{k=0}^{Q+m} b_k b_{k-m}^{\langle \alpha-1 \rangle} \right) & \text{for } -Q \leq m \leq 0 \\ 0 & \text{for } |m| > Q. \end{cases}$$ (7.17)

This is a set of highly nonlinear equations. However, for $MA(1), MA(2)$, and $MA(3)$ processes, the closed-form solutions of b_k in terms of the output covariation have been found [Ma and Nikias, 1995]. Without loss of generality, assuming the input U is a standard $S\alpha S$ process, we have for $MA(1)$,

$$b_0 = \left([X_n, X_{n+1}]_\alpha \cdot [X_n, X_{n-1}]_\alpha^{\langle 1-\alpha \rangle} \right)^{1/(2\alpha - \alpha^2)}, \alpha \neq 2, \qquad (7.18)$$

$$b_1 = [X_n, X_{n-1}]_\alpha \cdot b_0^{\langle 1-\alpha \rangle}. \qquad (7.19)$$

For $MA(2)$,

$$b_0 = \left([X_n, X_{n+2}]_\alpha \cdot [X_n, X_{n-2}]_\alpha^{\langle 1-\alpha \rangle} \right)^{1/(2\alpha - \alpha^2)}, \alpha \neq 2, \qquad (7.20)$$

$$b_1 = \frac{\dfrac{[X_n, X_{n-1}]_\alpha}{b_2} - \dfrac{[X_n, X_{n+1}]_\alpha}{b_0}}{\dfrac{[X_n, X_{n-2}]_\alpha}{b_2^2} - \dfrac{[X_n, X_{n+2}]_\alpha}{b_0^2}}, \qquad (7.21)$$

$$b_2 = [X_n, X_{n-2}]_\alpha \cdot b_0^{\langle 1-\alpha \rangle}. \qquad (7.22)$$

7.3 PARAMETER ESTIMATION OF ARMA STABLE PROCESSES

It is also possible to obtain a closed-form solution for $MA(3)$ processes (see Problem 7). The above derivations show that it is possible to perform blind identification of a FIR channel when the input is a non-Gaussian $S\alpha S$ ($\alpha \neq 2$) process.

7.3.2 Sampling Results and Performance Comparisons

To study the large sample properties of the high-order Yule-Walker equation for the AR parameters of the ARMA model, we generated 5000 samples of the stationary $S\alpha S$ output of a second-order ARMA(2,2) model defined by

$$X(n) - a_1 X(n-1) - a_2 X(n-2) = U(n) + b_1 U(n-1) + b_2 U(n-2),$$

where

$$a_1 = 0.195, \quad a_2 = -0.95, \, b_1 = 0.91, \, b_2 = 0.009.$$

The AR parameters were estimated by using FLOM and SCR estimators for the covariation coefficients and by solving (7.16). The results obtained are illustrated in Table 7.2. From this table, it is clear that the high-order Yule-Walker method based

TABLE 7.2 Estimation of AR Parameters of ARMA(2,2) α-Stable Process

Coefficients	α	Yule-Walker(FLOM)	Yule-Walker(SCR)
$a_1 = 0.195$	1.1	0.1963 (0.1324)	0.1920 (4.4582)
	1.3	0.1957 (0.0804)	0.2270 (1.7746)
	1.5	0.1954 (0.05953)	0.1921 (0.7565)
	1.9	0.1946 (0.0449)	0.1838 (0.3143)
	2.0	0.1947 (0.0425)	0.1900 (0.2211)
$a_2 = -0.95$	1.1	−0.9524 (0.1117)	−0.9951 (11.0807)
	1.3	−0.9492 (0.0925)	−0.8979 (3.4833)
	1.5	−0.9519 (0.0643)	−0.9760 (1.0251)
	1.9	−0.9498 (0.0394)	−0.9479 (0.2210)
	2.0	−0.9505 (0.0336)	−0.95 (0.1939)

on FLOM estimates with $p = 1$ is superior to that based on the SCR methods in terms of accuracy and consistency.

7.4 OVERDETERMINED YULE-WALKER EQUATION AND SINGULAR VALUE DECOMPOSITION

As we have seen in the previous sections, when the number of samples is large, the AR estimates based on generalized or higher-order Yule-Walker equations are good. If only short data are available, the performance of Yule-Walker–based methods will deteriorate due to the inaccuracy of the estimated covariation coefficients. In this case, overdetermined Yule-Walker equations should be used and solved by principal component methods based on singular value decomposition (SVD) algorithm [Marple, 1987].

As an example of the improvement of AR estimates that may be obtained by using SVD, we generated 500 samples of the same stationary $S\alpha S$ output of a

TABLE 7.3 Estimation of Parameters of a Second-Order AR α-Stable Process with SVD Algorithm

Coefficients	α	Yule-Walker (FLOM)	Yule-Walker (SCR)	Yule-Walker (FLOM-SVD)	Yule-Walker (SCR-SVD)
	1.1	0.1934 (0.1740)	−0.1333 (10.5890)	0.1981 (0.0232)	0.1843 (0.1412)
	1.3	0.1963 (0.1473)	0.1442 (1.5819)	0.1976 (0.0219)	0.1595 (0.1213)
$a_1 = 0.195$	1.5	0.1928 (0.1492)	0.2041 (1.24981)	0.1925 (0.0203)	0.1761 (0.0837)
	1.9	0.2009 (0.1481)	0.2067 (0.7896)	0.2030 (0.0189)	0.2093 (0.0444)
	2.0	0.196614 (0.1197)	0.1967 (0.6347)	0.1996 (0.0138)	0.2027 (0.0489)
	1.1	−0.9494 (0.1872)	−2.21679 (42.0043)	−0.9646 (0.0193)	−0.9382 (0.2852)
	1.3	−0.9483 (0.1551)	−1.0027 (1.4040)	−0.9638 (0.0139)	−0.9595 (0.0941)
$a_2 = -0.95$	1.5	−0.9460 (0.1585)	−0.9725 (1.0569)	−0.9626 (0.0151)	−0.9409 (0.1030)
	1.9	−0.9517 (0.1276)	−0.9606 (0.6731)	−0.9696 (0.0121)	−0.9599 (0.0437)
	2.0	−0.9473 (0.1227)	−0.9584 (0.6208)	−0.9705 (0.0097)	−0.9688 (0.0366)

7.4 OVERDETERMINED YULE-WALKER EQUATION

second-order AR system defined by

$$X(n) - a_1 X(n-1) - a_2 X(n-2) = U(n),$$

where

$$a_1 = 0.195, \quad a_2 = -0.95.$$

Note that this time we use only 500 samples instead of 5000 samples as in Section 7.2.4. In the SVD procedure, ten equations were used instead of two. The SVD of the covariation coefficient matrix was performed and only the largest two components were retained. The generalized inverse was used to solve the overdetermined set of equations. The results obtained are summarized in Table 7.3. From this table, it is obvious that the SVD-based approach performs much better in terms of accuracy and stability.

To show the stability of these estimators more clearly, the corresponding AR model frequency responses with the estimated parameters are shown in Figures 7.6–7.10.

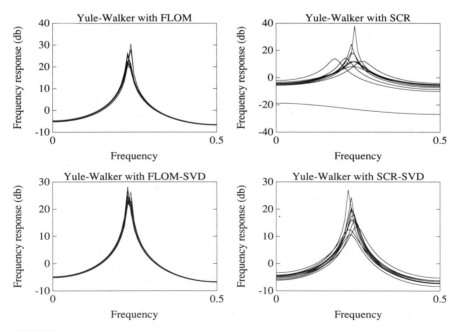

FIGURE 7.6 AR model frequency response with the estimated parameters when $\alpha = 1.1$.

92 PARAMETRIC MODELS OF STABLE PROCESSES

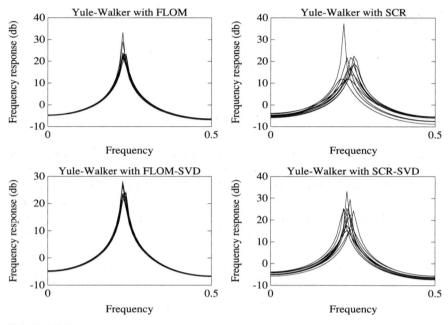

FIGURE 7.7 AR model frequency response with the estimated parameters when $\alpha = 1.3$.

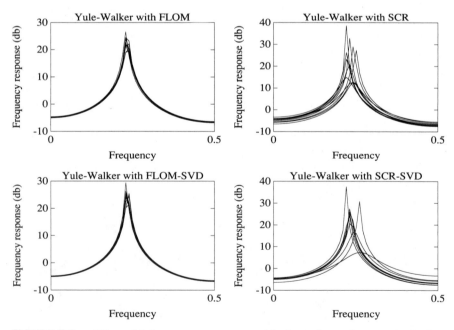

FIGURE 7.8 AR model frequency response with the estimated parameters when $\alpha = 1.5$.

7.4 OVERDETERMINED YULE-WALKER EQUATION

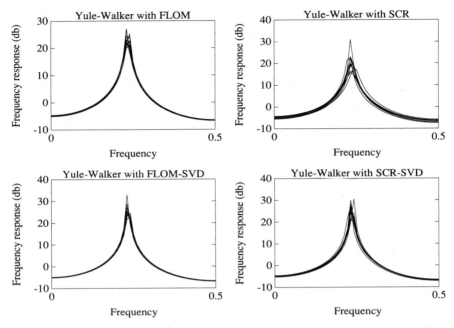

FIGURE 7.9 AR model frequency response with the estimated parameters when $\alpha = 1.9$.

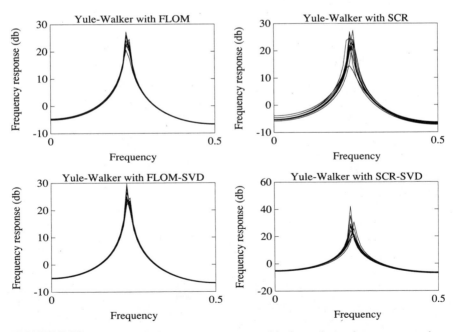

FIGURE 7.10 AR model frequency response with the estimated parameters when $\alpha = 2.0$.

The same SVD approach can be applied to the higher-order Yule-Walker equation for the ARMA model (see (7.16)).

7.5 THE YULE-WALKER EQUATION WITH MOMENTS OF ARBITRARY ORDER

The idea of using FLOMs in the Yule-Walker equation instead of second-order moments can be applied to any wide sense stationary AR process with finite moments of appropriate orders. Specifically, let us consider processes generated by the following AR system

$$X(n) = a_1 X(n-1) + \cdots + a_P X(n-P) + U(n) \tag{7.23}$$

where $\{U(n)\}$ is a sequence of uncorrelated random variables with finite moments of arbitrary order and zero mean. Assume all the AR model poles are inside the unit circle.

It is easy to show that if we define

$$\mathbf{p} = \begin{bmatrix} \lambda(1) \\ \lambda(2) \\ \vdots \\ \lambda(P) \end{bmatrix}, \quad \mathbf{a} = \begin{bmatrix} a_1 \\ a_2 \\ \vdots \\ a_P \end{bmatrix}$$

and

$$\mathbf{C} = \begin{bmatrix} \lambda(0) & \lambda(-1) & \cdots & \lambda(1-P) \\ \lambda(1) & \lambda(0) & \cdots & \lambda(2-P) \\ \vdots & \vdots & \ddots & \vdots \\ \lambda(P-1) & \lambda(P-2) & \cdots & \lambda(0) \end{bmatrix} \tag{7.24}$$

then the coefficients of the AR system can be found by solving the following system of Yule-Walker linear equations:

$$\mathbf{Ca} = \mathbf{p} \tag{7.25}$$

where

$$\lambda(k) = \mathbf{E}(X(n)X(n-k)^{\langle P-1 \rangle})/\mathbf{E}|X(n-k)|^p$$

and $p \geq 1$ is arbitrary.

To compare the performances of the above Yule-Walker estimator of order p with the LS and LAD estimators, we generated 5000 samples of the stationary

TABLE 7.4 Estimation of Coefficients of Second-Order AR Process

Estimator	a_1	a_2
Least-Squares	0.1952 (0.0297)	−0.9492 (0.03340)
LAD	0.1950 (0.0472)	−0.9491 (0.0576)
Yule-Walker ($p = 1$)	0.1943 (0.0383)	−0.9491 (0.0397)
Yule-Walker ($p = 2$)	0.1953 (0.0294)	−0.9492 (0.0340)
Yule-Walker ($p = 3$)	0.1954 (0.0311)	−0.9489 (0.0328)
Yule-Walker ($p = 4$)	0.1952 (0.0363)	−0.9487 (0.0357)

output of a second-order AR model defined by

$$X(n) - a_1 X(n-1) - a_2 X(n-2) = U(n)$$

where $U(n)$ is an i.i.d. sequence with uniform distribution and

$$a_1 = 0.195, \quad a_2 = -0.95.$$

The results obtained are presented in Table 7.4 and Figure 7.11. Clearly, all these methods perform very well.

7.6 CONCLUSION

In this chapter, we introduced methods for the estimation of AR and ARMA model parameters based on the covariation coefficients of the output observations. The generalized Yule-Walker equation approach was described, along with the least-squares method and the least absolute deviation criterion. Sampling results from several examples and performance comparisons among these methods were presented. Finally, we described how the SVD algorithm can be applied on covariation coefficient matrices to obtain more stable solutions of the generalized Yule-Walker equation.

PROBLEMS

1. Let us consider the following AR system:

$$X(n) = a_1 X(n-1) + a_2 X(n-2) + \cdots + a_P X(n-P) + U(n),$$

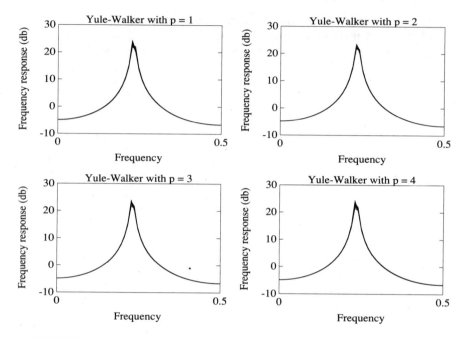

FIGURE 7.11 AR model frequency response with the estimated parameters.

where the $U(n)$'s are i.i.d. $S\alpha S$ random variables with characteristic exponent α and dispersion γ_u. The output $X(n)$ is also a $S\alpha S$ random variable. Calculate the output dispersion γ_x.

2. (Continuing Problem 1) For the AR model driven by $S\alpha S(1 < \alpha \leq 2)$ processes, there are three consistent estimators for its coefficients: (i) least squares, (ii) least absolute deviation, and (iii) generalized Yule-Walker based on FLOM covariations. Develop appropriate estimator(s) for $\alpha \leq 1$. (*Hint*: Covariation is undefined when $\alpha \leq 1$.)

3. (Continuing Problem 1) Suppose the input $U(n)$ is a sequence of i.i.d. $S\alpha S$ random variables with unknown characteristic exponent α and unit dispersion γ_u. How do you estimate the output characteristic exponent α and dispersion γ_x from $X(n)$ (output data) only?

4. The least squares estimator proposed by Kanter and Steiger can be used to estimate the coefficients of any AR model and the performance is fairly good even when the model is driven by a non-Gaussian $S\alpha S$ process. The only necessary condition is that the AR model is stable, i.e., all poles are inside the unit circle. Kanter and Steiger also applied this algorithm to estimate the coefficients of the MA model:

$$X_n = \sum_{i=0}^{q} c_i U_{n-i},$$

where c_n are the coefficients and U_n's are i.i.d. $S\alpha S$ random variables. However, this approach only works for minimum-phase MA models (all zeros are inside the unit circle). What happens if we apply this method to a nonminimum-phase MA model?

5. In general, for an AR model, it is relatively easy to estimate the coefficients from the output (generalized Yule-Walker method). For ARMA models, it is possible to break the problem into AR and MA problems. Therefore, the key problem is how to estimate the MA model coefficients. However, for the MA model, we have to solve a system of nonlinear equations (Sec. 7.3.1) in order to obtain the coefficients. How do you solve this system of nonlinear equations? Implement your solutions by Monte Carlo simulations and compare their performance.

6. By the definition of codifference (Problem 8, Chapter 6), we can also obtain a critically determined system of nonlinear equations by taking codifference of output samples with different time lags. The advantage of codifference is that it is defined for $\forall \alpha \in (0, 2]$. Use the least squares method to solve the system of nonlinear equations to obtain the MA model coefficients.

7. (Continuing Problem 6) In Problem 6, we used the least squares method to solve the system of nonlinear equations. However, it is found that the solution is not unique, i.e., when we start with different initial guesses, the final solutions may be different. As a matter of fact, the codifference approach is phase blind. Consider the following two FIR channels that have the same magnitude response but different phase response:

Minimum-phase:

$$H(z) = (1 - az^{-1})(1 - bz^{-1}), \quad |a| < 1, |b| < 1.$$

The coefficients are $h_0 = 1, h_1 = -a - b, h_2 = ab$, and

Maximum-phase:

$$H(z) = (z^{-1} - a)(z^{-1} - b).$$

The coefficients are $h_0 = ab, h_1 = -a - b, h_2 = 1$. Show that the system of codifference equations for these two channels is the same.

8. We have shown that it is possible to obtain a closed-form solution of the coefficients of some simple FIR channels. Find the closed-form solution of the following FIR system:

$$Y_n = h_0 X_n + h_1 X_{n-1} + h_2 X_{n-2} + h_3 X_{n-3}.$$

Run Monte Carlo simulations when: (1) the characteristic exponent α and the output dispersion γ_y are known exactly, (2) these parameters are not known and have to be estimated from the output.

8

Linear Theory of Stable Processes

8.1 INTRODUCTION

One of the central problems in statistical signal processing can be stated as follows: given a set of observations $\{X(t), t \in T\}$, find the "best" estimate of an unknown random variable Y from the linear space spanned by $\{X(t), t \in T\}$. This is the so-called *linear theory* of stochastic processes, which includes linear estimation, prediction, and filtering.

The linear theory of second-order processes (Gaussian processes in particular) has been fully developed. In this case, the linear space $L(X(t), t \in T)$ of the observations $\{X(t), t \in T\}$ is a Hilbert space. Under the *minimum mean-squared error* (MMSE) criterion, the best linear estimate of the unknown Y is the orthogonal projection of Y onto $L(X(t), t \in T)$. On the other hand, the linear theory of non-Gaussian stable processes has only recently been the subject of intensive research and relatively few results are available. A major difficulty is that the linear span of a stable process is a Banach space when $1 \le \alpha < 2$ and only a metric space when $0 < \alpha < 1$. These spaces do not have as nice properties and structures as Hilbert spaces for the linear estimation problem.

Although the development of linear theory for general stable processes is still at primitive stages, there are explicit results for special types of stable processes, such as harmonizable, linear, and sub-Gaussian processes. A large proportion of these results are devoted to harmonizable stable processes [Pourahmad, 1984; Weron, 1985; Cambanis and Miamee, 1989]. They are usually presented in the spectral domain and cannot be easily implemented as signal processing algorithms. In this chapter, we focus our attention on the prediction and estimation of discrete-time linear processes.

8.2 THE MINIMUM DISPERSION CRITERION

Let us first look at what we mean by a "best" estimate. For second-order processes, the most commonly used criterion for the best estimate is the MMSE criterion. Under this criterion, the best estimate is the one that minimizes the variance of the estimation error. If the process is Gaussian, it can be shown that this criterion also minimizes the probability of large estimation errors. For stable processes, the MMSE criterion is no longer appropriate due to the lack of finite variance. But the concept of MMSE criterion can be easily generalized to stable processes. Specifically, the *minimum dispersion* (MD) criterion is used in discussing linear theory of stable processes. Under the MD criterion, the best estimate of a $S\alpha S$ random variable in the linear space of observations is the one that minimizes the dispersion of the estimation error. Recall that the dispersion (i.e., the scale parameter) of a stable random variable plays a role analogous to the variance. For example, the larger the dispersion, the more the stable random variable spreads around its median. Thus, by minimizing the dispersion we minimize the average magnitude of estimation errors. Furthermore, it can be shown that minimizing the dispersion is also equivalent to minimizing the probability of large estimation errors [Cline and Brockwell, 1985]. The MD criterion is thus well justified in the stable case. It is a direct generalization of the MMSE criterion (they are the same in the Gaussian case) and reasonably simple to calculate. This criterion was introduced by Stuck [1978] in an attempt to solve a Kalman filtering type of problem associated with stable processes. It has also been used in Blattberg and Sargent [1971] for regression models with stable errors as well as in Cline and Brockwell [1985] for linear prediction of ARMA processes with infinite variance.

From (3.5), we see that the MD criterion is also equivalent to minimizing the FLOMs of estimation errors. These FLOMs measure the L_p distance between Y and its estimate $\hat{Y}$ on the linear space generated by the observation $\{X(t), t \in T\}$, for $p < \alpha$. This result is not surprising since the L_p norms for $p < 2$ are well known for their robustness against outliers such as those that may be described by stable laws. Notice that in the stable case all of the FLOMs are equivalent. A common choice is the L_1 norm, which is sometimes very convenient.

Under the MD criterion, the generic linear estimation problem of stable processes can be formulated as follows: Find an element $\hat{Y}$ in the linear space $L(X(t), t \in T)$ of the observations $\{X(t), t \in T\}$ such that

$$\|Y - \hat{Y}\|_\alpha = \inf_{Z \in L(X(t), t \in T)} \|Y - Z\|_\alpha \tag{8.1}$$

or equivalently

$$\mathbf{E}|Y - \hat{Y}|^p = \inf_{Z \in L(X(t), t \in T)} \mathbf{E}|Y - Z|^p \tag{8.2}$$

for $0 < p < \alpha$. Since $L(X(t), t \in T)$ is a Banach space, $\hat{Y}$ always exists and is unique for $1 < \alpha < 2$ [Singer, 1970]. It is obtained by a metric projection of Y

onto the convex Banach space $L(X(t), t \in T)$. For $1 < \alpha < 2$, $\hat{Y}$ is also uniquely determined [Cambanis and Miller, 1981] by

$$[X(t), Y - \hat{Y}]_\alpha = 0 \quad \text{for all } t \in T. \tag{8.3}$$

This is analogous to the *orthogonality principle* used extensively in the linear estimation problem of second-order processes [Papoulis, 1991].

When $\alpha = 2$, (8.3) is linear and thus a closed-form solution exists for $\hat{Y}$. For $\alpha < 2$, it is highly nonlinear and hard to solve for the estimate $\hat{Y}$. For example, let X, Y be jointly $S\alpha S$ with spectral measure $\mu(\cdot)$. Let aY be the optimal approximation of X in the MD sense. Then, the coefficient a is the solution of the following equation

$$\int_S y(x - ay)^{\langle \alpha - 1 \rangle} \mu(d\mathbf{s}) = 0. \tag{8.4}$$

Even if we know $\mu(\cdot)$, the above equation is still very hard to solve.

A related estimation problem is the regression. The regression estimate of Y given $X_1, \ldots, X_n$ is the conditional expectation $\mathbf{E}(Y \mid X_1, \ldots, X_n)$. When $Y, X_1, \ldots, X_n$ are jointly Gaussian, the regression estimate is linear and is equal to the linear estimate determined by (8.3). When $Y, X_1, \ldots, X_n$ are jointly stable with $\alpha < 2$, this is no longer true, as we have seen in Theorem 7. Even if the regression estimate is linear, it need not be the same as the linear estimate. A simple example to illustrate the point is to consider the regression of two $S\alpha S$ random variables X and Y. In this case, the regression is given by $\mathbf{E}(Y \mid X) = aX$, where $a = [Y, X]_\alpha / [Y, Y]_\alpha$. On the other hand, the MD estimate $\hat{Y} = aX$ is determined by (8.4), which has no closed-form solutions.

8.3 A SUBOPTIMAL STATE-SPACE PREDICTION

Stuck [1978] attempted to extend the classic Kalman filtering theory to a more general situation where the plant and observation noises are stable processes. He considered only the scalar discrete-time case. Specifically, let the model of the process $X(n)$ and observation $Y(n)$ be given by

$$\begin{cases} X(n+1) = a\, X(n) + b\, U(n) \\ Y(n) = c\, X(n) + W(n) \end{cases} \tag{8.5}$$

where the plant noise $U(n)$ and observation noise $W(n)$ are two i.i.d. $S\alpha S$ sequences with dispersions γ_u and γ_w, respectively. In addition, we assume that $U(n), W(m)$ are independent for all n and m. The initial value $X(0)$ is also assumed to be stable with dispersion γ_{x_0}. To predict $X(n+1)$ from the observations $Y(n), Y(n-1), \ldots, Y(1)$, Stuck [1978] suggested using the recursive formulation of Kalman filters

$$\begin{aligned} \hat{X}(n+1) &= a\, \hat{X}(n) + G_n\, e(n) \\ e(n) &= Y(n) - c\, \hat{X}(n) \end{aligned} \tag{8.6}$$

where the gain G_n is chosen to minimize the dispersion $\gamma_{e(n)}$ of the prediction error $e(n)$. Note that $e(0) = X(0)$; thus $\gamma_e(0) = \gamma_{x_0}$. It can be shown [Stuck, 1978] that for $1 < \alpha \leq 2$

$$\begin{cases} G_n = \dfrac{a}{c} \dfrac{[|c|^\alpha \gamma_e(n)]^{1/(\alpha-1)}}{[|c|^\alpha \gamma_e(n)]^{1/(\alpha-1)} + \gamma_w^{1/(\alpha-1)}} \\ \gamma_e(n+1) = |b|^\alpha \gamma_u + \left|\dfrac{a}{c}\right|^\alpha \dfrac{\gamma_w |C|^\alpha \gamma_e(n)}{\{[|c|^\alpha \gamma_e(n)]^{1/(\alpha-1)} + \gamma_w^{1/(\alpha-1)}\}^{\alpha-1}} \end{cases} \quad (8.7)$$

Although (8.6) is easy to implement, it is only suboptimal. It does not truly use the MD criterion in the prediction, as it can be seen from the fixed recursive formulation. In general, the MD prediction of $X(n)$ no longer has the recursive formulation that the Kalman filter has for the Gaussian case.

8.4 MD PREDICTION OF LINEAR STABLE PROCESSES

Consider the ARMA stable process $X(n)$ determined by the following equation

$$X(n) - a_1 X(n-1) - \cdots - a_P X(n-P) = U(n) + b_1 U(n-1) + \cdots + b_Q U(n-Q) \quad (8.8)$$

where $\{U(n), n = 0, \pm 1, \pm 2, \ldots\}$ are i.i.d. $S\alpha S$ random variables. Let

$$\frac{1 - a_1 z^{-1} - \cdots - a_P z^{-P}}{1 + b_1 z^{-1} + \cdots + b_Q z^{-Q}} = 1 - \sum_{k=1}^{\infty} h_k z^{-k} \quad \text{for } |z| = 1 \quad (8.9)$$

be the inverse of the system. Then, the solution of k-step MD prediction based on the infinite past is given by the following theorem [Cline and Brockwell, 1985]:

Theorem 14 *For the ARMA(P, Q) process there exists a unique minimum dispersion linear predictor $\hat{X}(n+k)$ for $X(n+k)$, $k \geq 1$, based on the infinite past $X(n)$, $X(n-1), \ldots$, given by the following recursive relationship:*

$$\hat{X}(n+k) = \sum_{j=1}^{k-1} h_j \hat{X}(n+k-j) + \sum_{j=k}^{\infty} h_j X(n+k-j). \quad (8.10)$$

In practice, one would like to predict $X(n+k)$ based on finite observations $X(n), \ldots, X(1)$. In this case, the truncation of the solution in (8.10) gives a nearly optimal solution for large n. The exact MD k-step MD predictor of a general ARMA(P, Q) process based on the finite past $X(n), \ldots, X(1)$ is analytically involved, except for the AR process. In this case, one has [Cline and Brockwell, 1985]

Theorem 15 *For the AR process given by the following stable system*

$$X(n) = a_1 X(n-1) + \cdots + a_P X(n-P) + U(n)$$

where $U(n)$ is white stable noise, there exists a unique minimum dispersion k-step MD predictor $\hat{X}(n+k)$ for $X(n+k)$ ($k \geq 1, n \geq p$) in terms of $X(1), \ldots, X(n)$. It satisfies the following recursive relationship

$$\hat{X}(n+k) = a_1 \hat{X}(n+k-1) + \cdots + a_P \hat{X}(n+k-P) \tag{8.11}$$

with initial conditions $\hat{X}(j) = X(j)$ for $1 \leq j \leq n$.

Thus, the MD predictor $\hat{X}(n+k)$ is exactly the same as the MMSE predictor for an AR Gaussian process. This is not the case for general ARMA stable processes.

8.5 ADAPTIVE WIENER-TYPE FILTERS FOR STABLE PROCESSES

Adaptive solutions of linear estimation problems for stable processes are much easier to implement because they do not require closed-form expressions. The dispersion of the estimation error is usually a convex function of the parameters. So, numerical methods, such as stochastic gradient methods, may be used to find the parameters by minimizing the dispersion of the error function.

Let us consider designing an FIR filter with an input consisting of a stationary $S\alpha S$ process $\{u(0), u(1), u(2), \ldots\}$. The problem is to choose the tap weight $\{w_0, w_1, \ldots, w_{M-1}\}$ such that the output of the filter is as close to a given desired response $\{d(n)\}$ as possible. Here we assume $\{d(n)\}$ and $\{u(n)\}$ are jointly $S\alpha S$. Specifically, we would like to find $\{w_0, w_1, \ldots, w_{M-1}\}$ such that the dispersion of the error

$$e(n) = d(n) - \sum_{k=0}^{M-1} w_k u(n-k) \tag{8.12}$$

is minimized. The cost function is thus given by

$$J = \left\| d(n) - \sum_{k=0}^{M-1} w_k u(n-k) \right\|_\alpha . \tag{8.13}$$

This cost function turns out to be quite intractable in general. We will use an equivalent form. By Proposition 2, the norm of a $S\alpha S$ random variable is proportional to its pth order moment for any $0 < p < \alpha$. So, an equivalent cost function is given by

$$J = \mathbf{E}\left(\left| d(n) - \sum_{k=0}^{M-1} w_k u(n-k) \right|^p \right) \tag{8.14}$$

where $0 < p < \alpha$. A particularly simple case is when $p = 1$. In this case, the cost function is just

8.5 ADAPTIVE WIENER-TYPE FILTERS FOR STABLE PROCESSES

$$J = \mathbf{E}\left(\left|d(n) - \sum_{k=0}^{M-1} w_k u(n-k)\right|\right). \qquad (8.15)$$

There is no closed-form solution for the set of coefficients that minimize the cost function J in (8.14). But, J is convex for $1 \leq p < \alpha$ and so we may use a stochastic gradient method to solve for the coefficients in a similar way as the LMS does for $p = 2$ (MMSE criterion). Hence, we propose the following LMP (least mean p-norm) algorithm.

LMP Algorithm: Fix p so that $1 \leq p < \alpha$.

1. Filter output:

$$y(n) = \hat{\mathbf{w}}^T(n)\mathbf{u}(n) \qquad (8.16)$$

where $\hat{\mathbf{w}}^T(n) = [\hat{w}_0, \ldots, \hat{w}_{M-1}]$, $\mathbf{u}^T(n) = [u(n), \ldots, u(n-M+1)]$.

2. Estimation error:

$$e(n) = d(n) - y(n). \qquad (8.17)$$

3. Tap weight adaptation:

$$\hat{\mathbf{w}}(n+1) = \hat{\mathbf{w}}(n) + \mu\, \mathbf{u}(n)|e(n)|^{p-1} \text{sign}(e(n)) \qquad (8.18)$$

where $\mu > 0$ is the step size.

When $p = 1$, the above algorithm will be called the LMAD (least mean absolute deviation) algorithm. The LMAD is actually the familiar *signed LMS* algorithm, although it is derived in a different context.

To compare the performance of the LMS and LMAD algorithms, we set up the following experiment that involves a first-order AR process. Consider an $AR(1)$ $S\alpha S$ process $\{u(n)\}$ described by the difference equation

$$u(n) = au(n-1) + v(n) \qquad (8.19)$$

where $a = 0.99$ is the parameter of the process, and $\{v(n)\}$ is a $S\alpha S$ white noise process of dispersion $\gamma_0 = 1$. Let the AR process in (8.19) reach steady-state before processing the data to ensure stationarity. To estimate the parameter a, we implemented adaptive predictors of order $M = 1$ using the LMAD and LMS algorithms. The LMAD algorithm for the weight adaptation can be written as

$$\hat{w}_{LMAD}(n+1) = \hat{w}_{LMAD}(n) + \mu u(n-1)\text{sign}(e(n))$$

$$e(n) = u(n) - \hat{w}_{LMAD}(n)u(n-1)$$

$$w_{LMAD}(0) = 0.0.$$

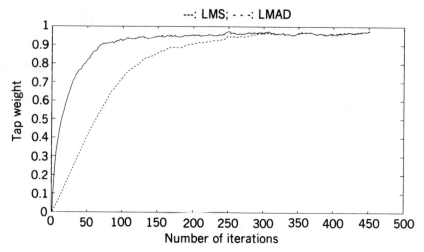

FIGURE 8.1 Transient behavior of tap weights in the LMS and LMAD algorithms with $\alpha = 2.0$ (Gaussian data).

On the other hand, the LMS can be written as

$$\hat{w}_{LMS}(n+1) = \hat{w}_{LMS}(n) + \mu u(n-1)e(n)$$

$$e(n) = u(n) - \hat{w}_{LMS}(n)u(n-1)$$

$$w_{LMS}(0) = 0.0.$$

Figures 8.1–8.4 show the plots of $\hat{w}_{LMAD}(n), \hat{w}_{LMS}(n)$ versus the number of iterations, where $\hat{w}_{LMAD}(n), \hat{w}_{LMS}(n)$ were obtained by averaging 30 independent trials

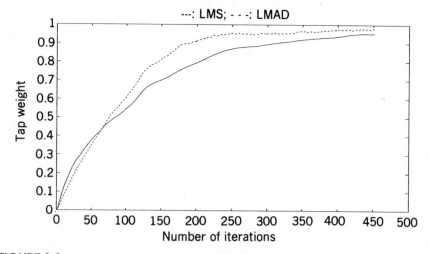

FIGURE 8.2 Transient behavior of tap weights in the LMS and LMAD algorithms with $\alpha = 1.9$.

8.5 ADAPTIVE WIENER-TYPE FILTERS FOR STABLE PROCESSES

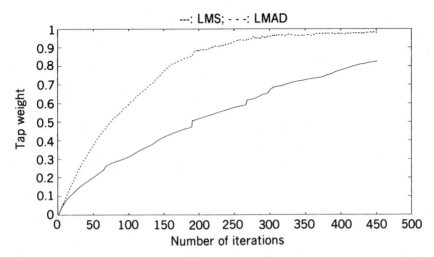

FIGURE 8.3 Transient behavior of tap weights in the LMS and LMAD algorithms with $\alpha = 1.5$.

of the experiment. For each trial, a different computer realization of the AR process $\{u(n)\}$ was used. To get a fair comparison between the LMAD and LMS regarding the convergence rates and misadjustments, in each experiment we first adjusted the step size of the LMS to the maximum while ensuring the convergence of the LMS and then chose a step size of the LMAD to get the same misadjustment as that of the LMS. Thus, both algorithms exhibit the same misadjustment (i.e., steady-state error) and their performances can be compared through the rate of convergence.

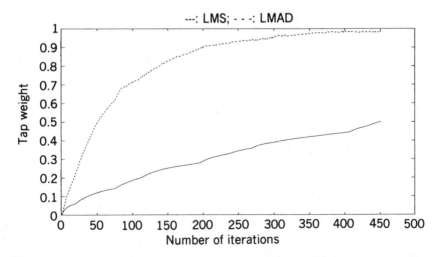

FIGURE 8.4 Transient behavior of tap weights in the LMS and LMAD algorithms with $\alpha = 1.1$.

A number of important observations can be made here. In all cases, the LMAD is simpler to implement than the LMS. In each iteration, the LMAD still needs M multiplications and $M-1$ additions to compute the filter output and prediction error, but the tap weight update equation is much simpler. This reduction in computation, however, comes at the expense of performance when the data are Gaussian. In a Gaussian environment ($\alpha = 2$), the LMAD is slower than the the LMS, as shown in Figure 8.1. This situation changes dramatically, however, when α is less than 2. As α decreases, the LMS becomes slower and slower to converge. In fact, when α is close to 1, the LMS hardly converges without its step size being very close to zero. On the other hand, the LMAD maintains fairly constant rate of convergence over the whole range of α. The LMAD converges faster and faster relative to the LMS as α becomes smaller and smaller while maintaining the same misadjustment as that of the LMS.

This experiment shows the distinct advantages of the LMAD when the AR process is driven by white stable noise. It is a simple, effective method for adaptive filtering when we are dealing with stable processes.

8.6 IDENTIFICATION OF LSI SYSTEMS

We consider a linear shift-invariant (LSI) system with a $S\alpha S$ process $X(n)$ as the input. Assume the impulse response of the system is $h(n)$. Then, the output $Y(n)$ is given by the following convolutional relationship

$$Y(n) = \sum_{k=0}^{\infty} h(k)X(n-k). \tag{8.20}$$

Observe that the output $Y(n)$ is also a $S\alpha S$ process and $Y(n)$ and $X(n)$ are jointly stationary.

In the following, we show that if we know the covariation sequence $C_{xx}(n)$ and the cross covariation sequence $C_{yx}(n)$, we can identify the system (that is, find $h(n)$). In fact, by taking covariations of both sides with $X(n-m)$, one has the following relation

$$C_{yx}(m) = \sum_{k=1}^{\infty} h(k)C_{xx}(m-k). \tag{8.21}$$

We now take the Fourier transform of both sides (assuming they exist) to get

$$\hat{C}_{yx}(\omega) = H(\omega)\hat{C}_{xx}(\omega) \tag{8.22}$$

where $\hat{C}_{yx}(\omega), H(\omega)$ and $\hat{C}_{xx}(\omega)$ are the Fourier transforms of $C_{yx}(m), h(n)$ and $C_{xx}(n)$, respectively.

So, in theory, one may recover the impulse response $h(n)$ and/or its Fourier transform $H(\omega)$ from the covariation sequences through

$$H(\omega) = \hat{C}_{yx}(\omega)/\hat{C}_{xx}(\omega). \tag{8.23}$$

Note, however, that $\hat{C}_{xx}(\omega)$ is just the Fourier transform of the covariation sequence of $X(n)$ and has no power spectrum meaning.

For a similar development for continuous-time systems, see Cambanis and Miller [1981].

8.7 CONCLUSION

In this chapter, attempts have been made to generalize the second-order linear theory for Gaussian processes to the lower-order linear theory for non-Gaussian stable processes. In particular, we have examined the role of minimum dispersion criterion and that of L_p norms with $1 \leq p < 2$ in linear prediction and filtering of $S\alpha S$ processes and the analysis and identification of linear systems with $S\alpha S$ inputs.

As we have seen in this chapter, the linear theory of stable processes is not simply a generalization of that of Gaussian processes despite the fact that stable processes are a one-step generalization of Gaussian processes. In general, closed-form solutions are not possible in the linear theory of stable processes. However, their adaptive implementation in signal processing is feasible and promising.

PROBLEMS

1. For an unknown random variable Y and its estimate $\hat{Y}$, show that the minimum dispersion criterion, minimum FLOM (minimize $E|Y - \hat{Y}|^p, 0 < p < \alpha$), and minimum norm (minimize $\|Y - \hat{Y}\|_\alpha$) are all equivalent.
2. The suboptimal state-space prediction algorithm proposed by Stuck was discussed briefly in Sec. 8.3. Derive the details of the extended Kalman filtering theory in $S\alpha S$ noise environment, then run Monte-Carlo simulations for (i) $\alpha = 2$ (Gaussian) and (ii) $1 < \alpha < 2$ (stable non-Gaussian). Compare the predicted results with their true values.
3. The LMP (least mean p-norm) algorithm is the basis for the adaptive Wiener-type filter of stable processes. The cost function J is a norm for $1 \leq p < \alpha$. When $p < 1$, the cost function J is not a norm. Repeat the experiment of Sec. 8.5 with different values of $p : 0 < p < \alpha$. Does ill-convergence occur when $p < 1$?
4. (Continuing Problem 3) When $\alpha \leq 1$, the cost function is not a norm since $0 < p < \alpha$. Does the LMP algorithm still work for $\alpha \leq 1$? If it does not, suggest an appropriate algorithm for this case.

5. Identification of a linear time-invariant system is a very important subject in signal processing. Is the system identifiable when (i) the input is a sequence of i.i.d. Gaussian random variables? (ii) the input is a sequence of i.i.d. non-Gaussian $S\alpha S$ random variables? If yes, present your identification methods; if no, explain the reason.
6. (Continuing Problem 5) A more challenging problem is to identify a LTI system from the output only (blind identification). Is the system *blindly* identifiable when (i) the input is a sequence of i.i.d. Gaussian random variables? (ii) the input is a sequence of i.i.d. non-Gaussian $S\alpha S$ random variables? If yes, present your identification methods; if no, explain the reason.

9

Symmetric Stable Models for Impulsive Noise

9.1 INTRODUCTION

Since the mid-1950s, there have been considerable efforts in developing accurate statistical models for non-Gaussian impulsive noise. The models that have been proposed so far may be roughly categorized into two groups: empirical models and statistical-physical models. Empirical models are the results of attempting to fit the experimental data by familiar mathematical expressions without considering physical grounds of the noise process. Commonly used empirical models include the hyperbolic distribution, Hall's generalized "t" distribution, and Gaussian mixture models [Mertz, 1961; Hall, 1966; Wegman, Schwartz, and Thomas, 1989].

Empirical models are usually simple and thus may lead to analytically tractable signal processing algorithms. However, they may not be motivated by the physical mechanism that generates the noise process. Hence, their parameters are often physically meaningless. In addition, applications of the empirical models are usually limited to particular situations.

Statistical-physical models, on the other hand, are derived from the underlying physical process giving rise to the noise, which takes into account the distribution of noise sources in space and time, and their propagations to the receiver [Furutsu and Ishida, 1961; Giordano and Haber, 1972; Middleton, 1977]. These models provide a realistic description of the underlying physics of noise processes and are known to fit closely a variety of non-Gaussian noises encountered in practice. However, they are usually mathematically involved.

In this chapter, we use symmetric stable distributions to model the first-order statistics of impulsive noise. It is shown that the stable model can be derived from the familiar filtered-impulse mechanism of the noise process under appropriate assumptions on the spatial and temporal distributions of noise sources and their propagation conditions. Compared with existing models, the stable model is much simpler and more accessible to engineers. In addition, it is found to be consistent with both experimental data and prior theoretical results.

9.2 FILTERED-IMPULSE MECHANISM OF NOISE PROCESSES

Most of the natural and man-made impulsive interferences may be considered as the results of a large number of spatially and temporally distributed sources that produce random noise pulses of short duration. The noise received at any location is the superposition of these pulses, and hence depends on the spatial and temporal distributions of the individual sources and the propagation of their pulses to the receiver. On the basis of these observations, the following filtered-impulse mechanism [Furutsu and Ishida, 1961; Middleton, 1977] is used in the development of the stable model for impulsive noise.

Let us assume, without loss of generality, that the origin of the spatial coordinate system is at the point of observation. The time axis is taken in the direction of past with its origin at the time of observation, i.e., t is the time length from the time of pulse occurrence to the time of observation. Consider a region Ω in R^n, where R^n may be a line ($n = 1$), a plane ($n = 2$), or the entire space ($n = 3$). For simplicity, we assume that Ω is a cone with vertex at the point of observation. Inside this region, there is a collection of noise sources (e.g., lighting discharges) that randomly generate transient pulses. It is assumed that all sources share a common random mechanism so that these elementary pulses have the same type of waveform, $aD(t; \underline{\theta})$, where the symbol $\underline{\theta}$ represents a collection of time-invariant random parameters that determine the scale, duration, etc., and a is a random amplitude. We shall further assume that only a countable number of such sources exist inside the region Ω, distributed at random positions $\mathbf{x}_1, \mathbf{x}_2, \ldots$. These sources emit pulses, $a_i D(t; \underline{\theta}_i), i = 1, 2, \ldots$, independently at random times $t_1, t_2, \ldots$, respectively. This implies that the random amplitudes $\{a_1, a_2, \ldots\}$ and the random parameters $\{\underline{\theta}_1, \underline{\theta}_2, \ldots\}$ are both i.i.d. sequences, with the prespecified probability densities $f_a(a)$ and $f_{\underline{\theta}}(\underline{\theta})$ respectively. The location $\mathbf{x}_i$ and emission time t_i of the ith source, the random parameter $\underline{\theta}_i$, and the amplitude a_i are assumed to be independent for every $i = 1, 2, \ldots$. Also, all of them are independent of all other random variables. The distribution of the random amplitude a, $f_a(a)$ is assumed to be symmetric, which in particular implies that the mean of the noise is zero.

When an elementary transient pulse $aD(t; \underline{\theta})$ passes through the medium and the receiver, it will be distorted and attenuated. The exact nature of distortion and attenuation can be determined from a knowledge of the beam patterns of source and antenna, source location, impulse response of receiver, etc., as Middleton [1974] has demonstrated. For simplicity, we shall assume that the effect of the transmission medium and the receiver on the transient pulses may be separated into two multiplicative factors: filtering and attenuation. Without attenuation, the medium and the receiver together may be treated as a deterministic linear time-invariant filter. In this case, the received transient pulse is the convolution of the impulse response of the equivalent filter and the original pulse waveform $aD(t; \underline{\theta})$. The result is designated by $aE(t; \underline{\theta})$.

The attenuation factor is generally a function of the source location relative to the receiver, and may be derived from the beam patterns of source and antenna and physical laws of electronic-magnetic transmission in the medium [Middleton,

9.2 FILTERED-IMPULSE MECHANISM OF NOISE PROCESSES

1974]. As a first-order approximation, we shall assume that the sources within the region of consideration have the same isotropic radiation pattern and the receiver has an omnidirectional antenna. Then, the attenuation factor is simply a decreasing function of the distance from the source to the receiver. A good approximation is that the attenuation factor varies inversely with a power of the distance [Giordano and Haber, 1972; Middleton, 1977], i.e.,

$$g(\mathbf{x}) = c_1/|\mathbf{x}|^\nu \tag{9.1}$$

where c_1 is a positive constant and $\nu > 0$ is the attenuation rate. Typically, $1/2 \leq \nu < 2$. For example, the waveguide-mode theory of long-distance propagation in the atmosphere shows that the attenuation factor is approximately given by (9.1), with $\nu = 1/2$ [Watt and Maxwell, 1957].

Combining the filtering and attenuation factors, one finds that the waveform of the received pulse originated from a source located at $\mathbf{x}$ is $aU(t; \mathbf{x}, \underline{\theta})$, where

$$U(t; \mathbf{x}, \underline{\theta}) = \frac{c_1}{|\mathbf{x}|^\nu} E(t; \underline{\theta}), \qquad c_1, \nu > 0. \tag{9.2}$$

Under the assumption that the receiver linearly superimposes the noise pulses, the observed instantaneous noise amplitude at the output of the receiver and at the time of observation is

$$X = \sum_{i=1}^{N} a_i\, U(t_i; \mathbf{x}_i, \underline{\theta}_i) \tag{9.3}$$

where N is the total number of noise pulses arriving at the receiver at the time of observation.

A basic assumption in our model as well as in many of the previous models for noise-generating processes [Furutsu and Ishida, 1961; Giordano and Haber, 1972; Middleton, 1974] is that the number of arriving pulses, N, is a Poisson point process in both space and time, whose intensity function is denoted by $\rho(\mathbf{x}, t)$. The intensity function $\rho(\mathbf{x}, t)$ represents the approximate probability that a noise pulse originated from a unit area or volume and emitted during a unit time interval will arrive at the receiver at the time of observation, and thus may be considered as the spatial and temporal density of the noise sources. Since the noise sources need not be homogeneously distributed in either space or time, the most general form of the intensity function $\rho(\mathbf{x}, t)$ depends on both the source location $\mathbf{x}$ and its emission time t, where $\mathbf{x} \in \Omega$ and $t \in R^+ = [0, \infty)$. However, for simplicity, we shall restrict our consideration to the most common case where the source distribution is time-invariant so that $\rho(\mathbf{x}, t) = \rho(\mathbf{x})$.

In most applications, $\rho(\mathbf{x})$ is a nonincreasing function of the range $|\mathbf{x}|$. The number of sources that occur close to the receiver is usually larger than the number of sources that occur farther away. This is certainly the case, for example, for the tropical atmospheric noise where most lighting discharges occur locally, and relatively few discharges occur at great distances [Giordano and Haber, 1972]. If the source distribution is symmetric about the point of observation, i.e., there is no preferred direction from which the pulses arrive, then it is reasonable to assume that $\rho(\mathbf{x})$ varies inversely with a certain power of the distance $|\mathbf{x}|$ [Middleton, 1977; Giordano and Haber, 1972]:

$$\rho(\mathbf{x}, t) = \frac{\rho_0}{|\mathbf{x}|^\mu} \qquad (9.4)$$

where $\mu, \rho_0 > 0$ are constants.

9.3 CHARACTERISTIC FUNCTION OF THE INSTANTANEOUS NOISE AMPLITUDE

Our method for calculating the characteristic function $\varphi(\omega)$ of the noise amplitude X is similar to the one used in Zolotarev [1986] for the model of point sources of influence. We first restrict our attention to noise pulses emitted from sources inside the region $\Omega(R_1, R_2)$ and within the time interval $[0, T)$, where $\Omega(R_1, R_2) = \Omega \cap \{\mathbf{x} : R_1 < |\mathbf{x}| < R_2\}$. The amplitude of the truncated noise is then given by

$$X_{T,R_1,R_2} = \sum_{i=1}^{N_{T,R_1,R_2}} a_i \, U(t_i; \mathbf{x}_i, \underline{\theta}_i), \qquad (9.5)$$

where N_{T,R_1,R_2} is the number of pulses emitted from the space-time region $\Omega(R_1, R_2) \times [0, T)$. The observed noise amplitude X is understood to be the limit of X_{T,R_1,R_2} as $T, R_2 \to \infty$ and $R_1 \to 0$ in some suitable sense.

Note that N_{T,R_1,R_2} is a Poisson variable with parameter

$$\lambda_{T,R_1,R_2} = \int_{\Omega(R_1,R_2)} \int_0^T \rho(\mathbf{x}, t) dt \, d\mathbf{x} \qquad (9.6)$$

and that its *factorial* moment-generating function is given by

$$E(t^{N_{T,R_1,R_2}}) = \exp(\lambda_{T,R_1,R_2}(t-1)). \qquad (9.7)$$

Let the actual source locations and their emission times be $(\mathbf{x}_i, t_i), i = 1, \ldots, N_{T,R_1,R_2}$. Then, the random pairs $(\mathbf{x_i}, t_i), i = 1, \ldots, N_{T,R_1,R_2}$ are independent and identically distributed, with a common joint density function given by

$$f_{T,R_1,R_2}(\mathbf{x}, t) = \frac{\rho(\mathbf{x}, t)}{\lambda_{T,R_1,R_2}}, \quad \mathbf{x} \in \Omega(R_1, R_2), \ t \in [0, T). \qquad (9.8)$$

In addition, N_{T,R_1,R_2} is independent of the locations and emission times of all the sources. All of the above results are the consequences of the basic Poisson assumption [Parzen, 1962].

9.3 CHARACTERISTIC FUNCTION OF THE INSTANTANEOUS NOISE AMPLITUDE

By our previous assumptions, $\{(a_i, t_i, \mathbf{x}_i, \underline{\theta}_i)\}_{i=1}^{\infty}$ is an i.i.d. sequence. Hence, X_{T,R_1,R_2} is a sum of i.i.d. random variables with a random number of terms. Its characteristic function can be calculated as follows:

$$\begin{aligned} \varphi_{T,R_1,R_2}(\omega) &= \mathbf{E}\{\exp(j\omega X_{T,R_1,R_2})\} \\ &= \mathbf{E}\{[\psi_{T,R_1,R_2}(\omega)]^{N_{T,R_1,R_2}}\}, \end{aligned} \qquad (9.9)$$

where

$$\psi_{T,R_1,R_2}(\omega) = \mathbf{E}\{\exp(j\omega a_1 U(t_1; \mathbf{x}_1, \underline{\theta}_1)) \mid T, \Omega(R_1, R_2)\}. \qquad (9.10)$$

By (9.7),

$$\varphi_{T,R_1,R_2}(\omega) = \exp(\lambda_{T,R_1,R_2}(\psi_{T,R_1,R_2}(\omega) - 1)). \qquad (9.11)$$

Since $a_1, \underline{\theta}_1$ and $(\mathbf{x}_1, t_1)$ are independent, with density functions $f_a(a), f_{\underline{\theta}}(\underline{\theta})$ and $f_{T,R_1,R_2}(\mathbf{x}, t)$, one obtains

$$\begin{aligned} \psi_{T,R_1,R_2}(\omega) = &\int_{-\infty}^{\infty} f_a(a) da \int_{\Theta} f_{\underline{\theta}}(\underline{\theta}) d\underline{\theta} \int_0^T dt \\ &\int_{\Omega(R_1,R_2)} \frac{\rho(\mathbf{x},t)}{\lambda_{T,R_1,R_2}} \exp(j\omega a U(t; \mathbf{x}, \underline{\theta})) d\mathbf{x}. \end{aligned} \qquad (9.12)$$

Combining (9.2), (9.4), (9.12), and (9.11), one can easily show that the logarithm of the characteristic function of X_{T,R_1,R_2} is

$$\begin{aligned} \log \varphi_{T,R_1,R_2}(\omega) = &\rho_0 \int_{-\infty}^{\infty} f_a(a) da \int_{\Theta} f_{\underline{\theta}}(\underline{\theta}) d\underline{\theta} \int_0^T dt \\ &\int_{\Omega(R_1,R_2)} [\exp(j\omega a c_1 |\mathbf{x}|^{-\nu} E(t; \underline{\theta})) - 1] |\mathbf{x}|^{-\mu} d\mathbf{x}. \end{aligned} \qquad (9.13)$$

Noting that the distribution of the random amplitude a is symmetric, one can further simplify the above equation to get

$$\begin{aligned} \log \varphi_{T,R_1,R_2}(\omega) = &2\rho_0 \int_0^{\infty} f_a(a) da \int_{\Theta} f_{\underline{\theta}}(\underline{\theta}) d\underline{\theta} \int_0^T dt \\ &\int_{\Omega(R_1,R_2)} [\cos(\omega a c_1 |\mathbf{x}|^{-\nu} E(t; \underline{\theta})) - 1] |\mathbf{x}|^{-\mu} d\mathbf{x}. \end{aligned} \qquad (9.14)$$

Now, $\log \varphi_{T,R_1,R_2}(\omega)$ approaches the following limit

$$\begin{aligned} \log \varphi(\omega) = &2\rho_0 \int_0^{\infty} f_a(a) da \int_{\Theta} f_{\underline{\theta}}(\underline{\theta}) d\underline{\theta} \int_0^{\infty} dt \\ &\int_{\Omega} [\cos(\omega a c_1 |\mathbf{x}|^{-\nu} E(t; \underline{\theta})) - 1] |\mathbf{x}|^{-\mu} d\mathbf{x} \end{aligned} \qquad (9.15)$$

as $T, R_2 \to \infty$ and $R_1 \to 0$, where $\varphi(\omega)$ is, of course, the desired characteristic function of the noise amplitude X.

To further simplify $\varphi(\omega)$, we rewrite the above integral involving the spatial coordinate $\mathbf{x}$ by using the polar coordinate system. Since Ω is a cone with vertex at the origin, (9.15) can be written as

$$\log \varphi(\omega) = 2c_2 \, \rho_0 \int_0^\infty f_a(a)\,da \int_\Theta f_{\underline{\theta}}(\underline{\theta})\,d\underline{\theta} \int_0^\infty dt$$
$$\int_0^\infty [\cos(\omega a c_1 |\mathbf{x}|^{-\nu} E(t;\underline{\theta})) - 1] |\mathbf{x}|^{\nu-1-\mu} dr \qquad (9.16)$$

where c_2 is a constant determined by the shape of Ω and n is the dimension of the space. We then make a change of variable, replacing $|\mathbf{x}|^{-\nu}$ by $|\mathbf{x}|$, to get

$$\log \varphi(\omega) = \frac{2}{\nu} c_2 \, \rho_0 \int_0^\infty f_a(a)\,da \int_\Theta f_{\underline{\theta}}(\underline{\theta})\,d\underline{\theta} \int_0^\infty dt$$
$$\int_0^\infty [\cos(\omega a c_1 |\mathbf{x}| E(t;\underline{\theta})) - 1] |\mathbf{x}|^{-\alpha-1} d|\mathbf{x}| \qquad (9.17)$$

where

$$\alpha = \frac{n-\mu}{\nu} \qquad (9.18)$$

is an effective measure of an average source density with range. As we will see in the next section, α determines the degree of impulsiveness of the noise.

We shall assume that $0 < \alpha < 2$. Otherwise, the integrand in (9.17) is not absolutely integrable. Using the following elementary integral [Gradshteyn and Ryzhik, 1965]

$$\int_0^\infty x^{\mu-1} \sin^2 ax\, dx = -\frac{\Gamma(\mu) \cos \frac{\mu \pi}{2}}{2^{\mu+1} a^\mu}, \quad a > 0,\ -2 < \mu < 0 \qquad (9.19)$$

one finally obtains the characteristic function of the instantaneous noise amplitude

$$\log \varphi(\omega) = -\gamma\, |\omega|^\alpha \qquad (9.20)$$

where

$$\gamma = C_1 \int_0^\infty a^\alpha f_a(a)\,da \int_\Theta f_{\underline{\theta}}(\underline{\theta})\,d\underline{\theta} \int_0^\infty dt\, |E(t;\underline{\theta})|^\alpha > 0 \qquad (9.21)$$

and where

$$C_1 = \frac{2 c_1^\alpha c_2 \rho_0 \Gamma(1-\alpha) \cos \frac{\pi}{2} \alpha}{\nu \alpha}.$$

Hence, we have shown that the probability distribution of the instantaneous amplitude of impulsive noise is symmetric stable ($S\alpha S$).

9.4 FIRST-ORDER STATISTICS OF ENVELOPE AND PHASE OF NARROWBAND NOISE

In the preceding section we have shown that the instantaneous noise amplitude is a $S\alpha S$ random variable. This result applies to both narrowband and wideband receptions. However, in most communication systems, the receiver is usually narrowband. In such cases, it is of interest to derive the first-order statistics of the envelope and phase of the received impulsive noise.

9.4.1 Joint Characteristic Function of the In-Phase and Quadrature Components

When the receiver is narrowband with central frequency ω_0, the typical waveform, $E(t; \underline{\theta})$, of the noise pulse after reception (see Sec. 9.2) may be represented by its slow-varying envelope $V(t; \underline{\theta})$ and random phase $\phi(t)$ as follows

$$E(t; \underline{\theta}) = V(t; \underline{\theta}) \cos(\omega_0 t + \phi(t)), \qquad (9.22)$$

where the envelope has a fixed waveform containing random parameters $\underline{\theta}$. Equivalently,

$$E(t; \underline{\theta}) = E_c(t; \underline{\theta}) \cos \omega_0 t - E_s(t; \underline{\theta}) \sin \omega_0 t \qquad (9.23)$$

where the in-phase and quadrature components are defined by

$$E_c(t; \underline{\theta}) = V(t; \underline{\theta}) \cos \phi(t), \quad E_s(t; \underline{\theta}) = V(t; \underline{\theta}) \sin \phi(t). \qquad (9.24)$$

By (9.3) and (9.2), the received narrowband noise at the time of observation is

$$X = X_c \cos \omega_0 t - X_s \sin \omega_0 t. \qquad (9.25)$$

Here the in-phase and quadrature components of the noise are given by

$$X_c = \sum_{i=1}^{N} a_i U_c(t_i; \mathbf{x}_i, \underline{\theta}_i, \phi_i), \quad X_s = \sum_{i=1}^{N} a_i U_s(t_i; \mathbf{x}_i, \underline{\theta}_i, \phi_i) \qquad (9.26)$$

where

$$U_c(t; \mathbf{x}, \underline{\theta}, \phi) = \frac{c_1}{|\mathbf{x}|^\nu} E_c(t; \underline{\theta}), \quad U_s(t; \mathbf{x}, \underline{\theta}, \phi) = \frac{c_1}{|\mathbf{x}|^\nu} E_s(t; \underline{\theta}) \qquad (9.27)$$

and $\phi_i = \phi(t_i)$ are the random phases of the elementary pulses at the time of observation. We shall assume that $\{\phi_i\}$ is an i.i.d. sequence, with uniform distribution in $[0, 2\pi]$ and independent of all other random variables.

To find the first-order statistics of the envelope and phase of the noise, we need the joint distribution of its quadrature components. In a similar way as in the derivation of (9.17) and taking into account the random phases, one can show that the joint characteristic function of X_c and X_s,

$$\varphi(\omega_1, \omega_2) = E(\exp(j(\omega_1 X_c + \omega_2 X_s))),$$

is given by

$$\log \varphi(\omega_1, \omega_2)) = \frac{1}{p\pi} c_2 p_0 \int_0^\infty f_a(a) da \int_\Theta f_{\underline{\theta}}(\underline{\theta}) d\underline{\theta} \int_0^\infty dt \int_0^{2\pi} d\phi \\ \int_0^\infty G_\alpha(a, \underline{\theta}, \phi, \lambda) d\lambda, \quad (9.28)$$

where the integrand is

$$G_\alpha(a, \underline{\theta}, \phi, \lambda) = [\cos(ac_1\lambda(\omega_1 E_c(t; \underline{\theta}) + \omega_2 E_s(t; \underline{\theta}))) - 1]\lambda^{-\alpha-1} \quad (9.29)$$

and α is again given by (9.18). Using (9.24), one can rewrite the integrand as follows

$$G_\alpha(a, \underline{\theta}, \phi, \lambda) = [\cos(ac_1 r|\underline{\omega}|V(t; \underline{\theta})\cos(\phi - \phi_\omega)) - 1]\lambda^{-\alpha-1}, \quad (9.30)$$

where $\underline{\omega} = (\omega_1, \omega_2)$, $|\underline{\omega}| = \sqrt{\omega_1^2 + \omega_2^2}$ and $\phi_{\underline{\omega}} = \arctan(\omega_2/\omega_1)$.

From (9.19), one finds that the joint characteristic function of the quadrature components of the noise is given by

$$\log \varphi(\omega_1, \omega_2)) = -\gamma |\underline{\omega}|^\alpha \quad (9.31)$$

where

$$\gamma = C_2 \int_0^\infty a^\alpha f_a(a) da \int_\Theta f_{\underline{\theta}}(\underline{\theta}) d\underline{\theta} \\ \int_0^\infty dt |V(t; \underline{\theta})|^\alpha \int_0^{2\pi} |\cos(\phi)|^\alpha d\phi > 0 \quad (9.32)$$

and where

$$C_2 = \frac{c_1^\alpha c_2 p_0 \Gamma(1-\alpha) \cos\frac{\pi}{2}\alpha}{\nu\alpha\pi}.$$

A bivariate random vector whose characteristic function $\varphi(\omega_1, \omega_2)$ has the form defined by (9.31) is called *isotropic α-stable*, which is a special case of multidimensional $S\alpha S$ random variables.

9.4.2 Bivariate Isotropic Stable Distributions

A bivariate isotropic α-stable distribution is, in its most general form, defined by a characteristic function of the form

$$\varphi(t_1, t_2) = \exp(j(a_1 t_1 + a_2 t_2) - \gamma |\underline{t}|^\alpha),$$
$$-\infty < a_1, a_2 < \infty, \quad \gamma > 0, \quad 0 < \alpha \leq 2 \tag{9.33}$$

where $\underline{t} = (t_1, t_2)$ and $|\underline{t}| = \sqrt{t_1^2 + t_2^2}$. The parameters a_1, a_2 determine its central location. The parameters γ and α are the dispersion and the characteristic exponent, respectively. Two important special cases are the bivariate isotropic Gaussian ($\alpha = 2$) and Cauchy ($\alpha = 1$) distributions. Note that the two marginal distributions of the isotropic stable distribution defined by (9.33) are $S\alpha S$ with parameters (a_1, γ, α) and (a_2, γ, α) respectively, a characteristic shared by general multivariate stable distributions.

As usual, we shall assume $(a_1, a_2) = (0, 0)$ so that the isotropic stable distribution is centered at the origin. Denote its density and distribution functions by $f_{\alpha,\gamma}(x_1, x_2)$ and $F_{\alpha,\gamma}(x_1, x_2)$, respectively. As in the univariate case, the density function $f_{\alpha,\gamma}(x_1, x_2)$ is the inverse Fourier transform of the characteristic function, i.e.,

$$f_{\alpha,\gamma}(x_1, x_2) = (2\pi)^{-2} \int_{t_1} \int_{t_2} \exp(-j(x_1 t_1 + x_2 t_2)) \exp(-\gamma |\underline{t}|^\alpha) dt_1 dt_2. \tag{9.34}$$

Let $\underline{x} = (x_1, x_2)$ and $|\underline{x}| = \sqrt{x_1^2 + x_2^2}$. Using the polar coordinate system for the integral, one can show that

$$f_{\alpha,\gamma}(x_1, x_2) = \chi_{\alpha,\gamma}(|\underline{x}|) \tag{9.35}$$

where

$$\chi_{\alpha,\gamma}(\lambda) = \frac{1}{2\pi} \int_0^\infty s \exp(-\gamma s^\alpha) J_0(s\lambda) ds, \quad r \geq 0 \tag{9.36}$$

and $J_n(x)$ is the nth-order Bessel function of the first kind.

It is apparent from the following integrals [Gradshteyn and Ryzhik, 1965]

$$\int_0^\infty z \exp(-az^2) J_0(bz) dz = \frac{1}{2a} \exp(-b^2/4a), \quad a > 0, b > 0$$

and

$$\int_0^\infty e^{-az} J_n(bz) z^{n+1} dz = \frac{a 2^{n+1} b^n \Gamma(n + 3/2)}{\sqrt{\pi}(a^2 + b^2)^{n+3/2}} \quad a > 0, b > 0, n > -1,$$

that

$$\chi_{2,\gamma}(\lambda) = \frac{1}{4\pi\gamma} \exp(-\lambda^2/4\gamma) \tag{9.37}$$

and

$$\chi_{1,\gamma}(\lambda) = \frac{\gamma}{2\pi(\lambda^2 + \gamma^2)^{3/2}}. \quad (9.38)$$

When $\alpha \neq 1$ or 2, no closed-form expression is known for the integral in (9.36). But, by exploiting the basic properties of Bessel functions, one can find the following series expansion [Zolotarev, 1981]:

$$\chi_{\alpha,\gamma}(\lambda) = \begin{cases} \dfrac{1}{\pi^2 \gamma^{2/\alpha}} \displaystyle\sum_{k=1}^{\infty} \dfrac{(-1)^{k-1} 2^{\alpha k}}{k!} (\Gamma(\alpha k/2 + 1))^2 \sin\left(\dfrac{k\alpha\pi}{2}\right) (\lambda/\gamma^{1/\alpha})^{-\alpha k - 2} \\ \qquad\qquad\qquad\qquad\qquad 0 < \alpha < 1 \\ \dfrac{1}{\pi \alpha \gamma^{2/\alpha}} \displaystyle\sum_{k=0}^{\infty} \dfrac{(-1)^k}{(k!)^2 2^{2k+1}} \Gamma\left(\dfrac{2k+2}{\alpha}\right) (\lambda/\gamma^{1/\alpha})^{2k} \\ \qquad\qquad\qquad\qquad\qquad 1 < \alpha < 2. \end{cases} \quad (9.39)$$

In addition, one can show that

$$\lim_{\lambda \to \infty} \lambda^{\alpha+2} \chi_{\alpha,\gamma}(\lambda) = B(\alpha, \gamma) \quad (9.40)$$

where $B(\alpha, \gamma)$ is a positive constant [Zolotarev, 1981]. Hence, $\chi_{\alpha,\gamma}(\lambda)$ has an algebraic tail of order $\alpha + 2$.

9.4.3 First-Order Statistics of Noise Envelope and Phase

From (9.25), the noise envelope A and phase Ψ are determined by the in-phase and quadrature components X_c and X_s as follows:

$$A = \sqrt{X_c^2 + X_s^2}, \quad \Psi = \arctan(X_s/X_c). \quad (9.41)$$

Since X_c, X_s are jointly isotropic α-stable with density function $f_{\alpha,\gamma}(x_1, x_2)$, the joint density of the envelope and phase is given by

$$f(a, \psi) = a f_{\alpha,\gamma}(a \cos\psi, a \sin\psi), \quad a \geq 0, \ 0 \leq \psi \leq 2\pi. \quad (9.42)$$

From (9.35) and (9.36), it follows that

$$f(a, \psi) = a \chi_{\alpha,\gamma}(a) = \frac{a}{2\pi} \int_0^\infty s \exp(-\gamma s^\alpha) J_0(sa) \, ds. \quad (9.43)$$

Since the bivariate density $f(a, \psi)$ is independent of ψ, the random phase is *uniformly distributed* in $[0, 2\pi]$ and is independent of the envelope. This is a well-known fact for the Gaussian case.

9.5 AMPLITUDE PROBABILITY DISTRIBUTION

The density of the envelope, on the other hand, is given by

$$f(a) = a \int_0^\infty s \exp(-\gamma s^\alpha) J_0(as) \, ds, \quad a \geq 0. \tag{9.44}$$

By integrating (9.44) and noting the following identity of Bessel functions [Gradshteyn, 1965]

$$\int_0^x s J_0(s) \, ds = x J_1(x),$$

one obtains the envelope distribution function

$$F(a) = a \int_0^\infty \exp(-\gamma t^\alpha) J_1(at) \, dt, \quad a \geq 0. \tag{9.45}$$

From (9.37), (9.38), and (9.39) the envelope density is also given by the following series

$$f(a) = \begin{cases} \dfrac{1}{\pi \gamma^{1/\alpha}} \displaystyle\sum_{k=1}^\infty \dfrac{(-1)^{k-1} 2^{\alpha k+1}}{k!} (\Gamma(\alpha k/2 + 1))^2 \sin\left(\dfrac{k\alpha\pi}{2}\right) \left(\dfrac{a}{\gamma^{1/\alpha}}\right)^{-\alpha k-1} & 0 < \alpha < 1 \\[2ex] \dfrac{a\gamma}{(a^2 + \gamma^2)^{3/2}} & \alpha = 1 \\[2ex] \dfrac{1}{\alpha \gamma^{1/\alpha}} \displaystyle\sum_{k=0}^\infty \dfrac{(-1)^k}{(k!)^2 2^{2k}} \Gamma\left(\dfrac{2k+2}{\alpha}\right) \left(\dfrac{a}{\gamma^{1/\alpha}}\right)^{2k+1} & 1 < \alpha < 2 \\[2ex] \dfrac{a}{2\gamma} \exp(-a^2/4\gamma) & \alpha = 2. \end{cases}$$

(9.46)

Note that when $\alpha = 2$, one obtains the familiar Rayleigh distribution for the envelope.

9.5 AMPLITUDE PROBABILITY DISTRIBUTION

A commonly measured statistical representation of impulsive noise is the amplitude probability distribution (APD), defined as the probability that the noise magnitude is above a threshold. Hence, if X is the instantaneous amplitude of impulsive noise, then its APD is given by $P(|X| > x)$ as a function of x. The APD can be easily measured in practice by counting the percentages of time for which the given threshold is exceeded by the noise magnitude during the period of observation.

In the case that X is $S\alpha S$ with dispersion γ, its APD can be calculated from (2.5) as follows

$$P(|X| > x) = 1 - \frac{2}{\pi} \int_0^\infty \frac{\sin tx}{t} \exp(-\gamma t^\alpha) dt. \tag{9.47}$$

By integrating (2.9), one can also represent the APD of $S\alpha S$ noise using power series as follows

$$P(|X| > x) = \begin{cases} \frac{2}{\pi\alpha} \sum_{k=1}^\infty \frac{(-1)^{k-1}}{k!k} \Gamma(\alpha k + 1) \sin\left(\frac{k\alpha\pi}{2}\right) \left(\frac{|x|}{\gamma^{1/\alpha}}\right)^{-\alpha k} & 0 < \alpha < 1 \\ 1 - \frac{2}{\pi} \arctan(x/\gamma) & \alpha = 1 \\ 1 - \frac{2}{\pi\alpha} \sum_{k=0}^\infty \frac{(-1)^k}{(2k+1)!} \Gamma\left(\frac{2k+1}{\alpha}\right) \left(\frac{x}{\gamma^{1/\alpha}}\right)^{2k+1} & 1 < \alpha < 2 \\ 1 - \text{erf}\left(\frac{x}{2\sqrt{\gamma}}\right) & \alpha = 2 \end{cases} \tag{9.48}$$

where erf(x) is the standard error function. Similarly, by integrating (2.20), one can easily show that

$$\lim_{x \to \infty} x^\alpha P(|X| > x) = (2/\alpha) C(\alpha, \gamma). \tag{9.49}$$

Hence, the APD of $S\alpha S$ noise decays asymptotically on the order of $x^{-\alpha}$. As we will see in Section 9.6, this result is consistent with experimental observations.

Figures 9.1 and 9.2 plot the APD of $S\alpha S$ noise for various values of α and γ. To fully represent the large range of the exceedance probability $P(|X| > x)$, the coordinate grid used in these two and subsequent figures employs a highly folded abscissa. Specifically, the axis for $P(|X| > x)$ is scaled according to $-\log(-\log P(|X| > x))$. As clearly shown in Figure 9.2, $S\alpha S$ distributions have a Gaussian behavior when the amplitude is below a certain threshold.

In the case of narrowband reception, the APD of the envelope is given by

$$P(A > a) = 1 - a \int_0^\infty \exp(-\gamma t^\alpha) J_1(at) dt, \quad a \geq 0. \tag{9.50}$$

9.5 AMPLITUDE PROBABILITY DISTRIBUTION

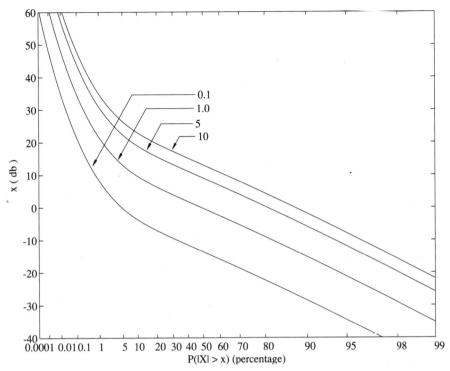

FIGURE 9.1 The APD of the instantaneous amplitude of $S\alpha S$ noise for $\alpha = 1.5$ and various values of γ.

By integrating (9.46), one obtains

$$P(A > a) = \begin{cases} \dfrac{1}{\pi\alpha} \sum_{k=1}^{\infty} \dfrac{(-1)^{k-1} 2^{\alpha k+1}}{k!k} (\Gamma(\alpha k/2 + 1))^2 \sin\left(\dfrac{k\alpha\pi}{2}\right) \left(\dfrac{a}{\gamma^{1/\alpha}}\right)^{-\alpha k} \\ \qquad\qquad\qquad\qquad\qquad\qquad 0 < \alpha < 1 \\[4pt] \dfrac{\gamma}{(a^2 + \gamma^2)^{1/2}} \qquad\qquad\qquad \alpha = 1 \\[4pt] 1 - \dfrac{1}{\alpha} \sum_{k=0}^{\infty} \dfrac{(-1)^k}{(k!)^2 (2k+2) 2^{2k}} \Gamma\left(\dfrac{2k+2}{\alpha}\right) \left(\dfrac{a}{\gamma^{1/\alpha}}\right)^{2k+2} \\ \qquad\qquad\qquad\qquad\qquad\qquad 1 < \alpha < 2 \\[4pt] \exp(-a^2/4\gamma) \qquad\qquad\qquad \alpha = 2. \end{cases} \qquad (9.51)$$

From (9.40), it follows that

$$\lim_{a \to \infty} a^{\alpha+1} f(a) = 2\pi B(\alpha, \gamma) > 0 \qquad (9.52)$$

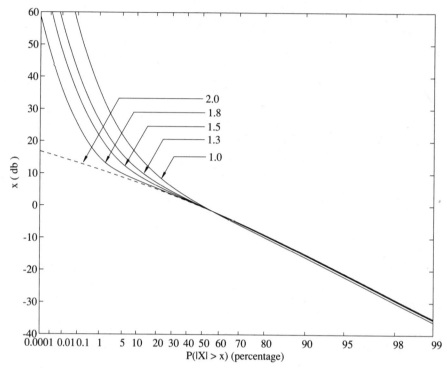

FIGURE 9.2 The APD of the instantaneous amplitude of $S\alpha S$ noise for $\gamma = 1$ and various values of α.

and hence

$$\lim_{a \to \infty} a^\alpha P(A > a) = (2\pi/\alpha)B(\alpha, \gamma), \qquad (9.53)$$

i.e., the envelope distribution and density functions are again heavy-tailed.

Figures 9.3 and 9.4 plot the APD of the $S\alpha S$ noise for various values of α and γ. Note that when $\alpha = 2$, i.e., when the envelope distribution is Rayleigh, one obtains a straight line with slope equal to $-\frac{1}{2}$. Figure 9.4 again shows that at low amplitudes the $S\alpha S$ noise is basically Gaussian (Rayleigh).

9.6 COMPARISONS WITH EXISTING MODELS AND EXPERIMENTAL DATA

Statistical-physical models are typically based on the same filtered-impulse mechanism used to derive the $S\alpha S$ model earlier in this paper, with different assumptions for the noise source distributions and propagation conditions [Furutsu and Ishida, 1961; Giordano and Haber, 1972; Middleton, 1977]. Among them, a particularly important model is Middleton's Class B model, which has been shown to fit a

9.6 COMPARISONS WITH EXISTING MODELS AND EXPERIMENTAL DATA

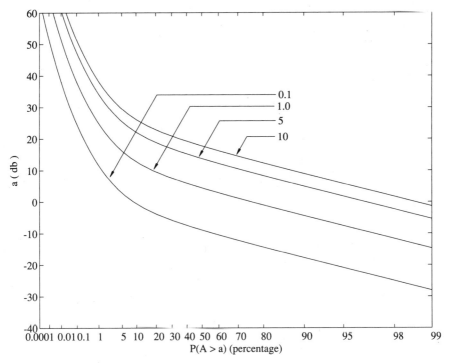

FIGURE 9.3 The APD of the envelope of narrowband $S\alpha S$ noise for $\alpha = 1.5$ and various values of γ.

variety of different types of impulsive noises, including atmospheric noise and automotive ignition noise.

In developing the Class B model, considerable attention is given to the detailed structures of the basic waveforms of the emissions and the spatial-temporal distribution of the noise sources [Middleton, 1977]. Unfortunately, this model is very complicated and difficult to derive. Furthermore, several approximations have to be made in its derivation to get usable results. The $S\alpha S$ model, on the other hand, is derived from simplified assumptions on the source density and propagation process. The main simplifications are that the beam patterns of the receiver antenna and noise sources are nondirectional and that the noise sources are isotropically distributed in space. However, the inverse-power characteristic of the propagation law and source density is preserved. No further approximation is used in the derivation, and the derivation is much simpler than that of the Class B model. It is interesting to note that Middleton's Class A model may be derived exactly if one uses similar simplifications for the source distributions and propagation conditions [Berry, 1981].

Despite the differences, the present $S\alpha S$ model and the Class B model are closely related. To see this, one notes that the characteristic function of the instantaneous

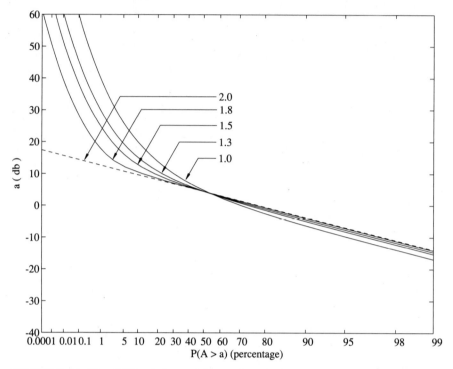

FIGURE 9.4 The APD of the envelope of narrowband $S\alpha S$ noise for $\gamma = 1$ and various values of α.

amplitude of a Class B interference in the absence of Gaussian background noise[1] can be approximated by two distinct functions that are suitably joined at some appropriate threshold z_{0B} [Middleton, 1978]. Specifically, when the noise amplitude is less than z_{0B}, the density function is approximately determined by the following characteristic function

$$\varphi_{B-I}(t) = \exp(-b_{1\alpha} A_B \hat{a}^\alpha |t|^\alpha). \tag{9.54}$$

When the noise amplitude is larger than z_{0B}, the approximate characteristic function is

$$\varphi_{B-II}(t) = e^{-A_B} \exp(A_B e^{-b_{2\alpha} \hat{a}^2 |t|^2 / 2}). \tag{9.55}$$

All the parameters in (9.54) and (9.55) are nonnegative and physically meaningful. In particular, the parameter α is exactly the same one defined by (9.18).

An immediate observation is that as the threshold z_{0B} approaches infinity, the $S\alpha S$ model and the Class B model coincide. Consequently, the $S\alpha S$ model may be

---

[1] Gaussian background noise is neglected in the development of the $S\alpha S$ model since its contribution is usually insignificant in the presence of impulsive noise.

9.6 COMPARISONS WITH EXISTING MODELS AND EXPERIMENTAL DATA

viewed as a limiting case of Middleton's Class B model. One can show that the tails of the distribution function determined by (9.55) are much lighter than the algebraic tails of $S\alpha S$ distributions. While this ensures that the Class B model has finite variance, it is often the algebraic behavior of the tails of $S\alpha S$ distributions that is of interest in practice, as we will demonstrate later. Also, the full Class B model is complicated and cannot be easily used in designing signal processing algorithms. This further suggests that the $S\alpha S$ model is more desirable than the full Class B model. One should also note that the envelope distributions of the $S\alpha S$ noise and Class B noise are generally different. While the characteristic function of the envelope of Class B noise can be approximated by two functions similar to those in (9.54) and (9.55) [Middleton, 1977], that of the envelope of $S\alpha S$ noise is a heavy-tailed generalization of the Rayleigh distribution.

We also observe that the stable model is consistent with empirical models. Many of the empirical models are based on the experimental observation that most of the impulsive noise, such as atmospheric noise, ice-cracking noise, and automotive ignition noise, are approximately Gaussian at low amplitudes and impulsive at high amplitudes [Crichlow et al., 1960; Wegman, Schwartz, and Thomas, 1989]. For example, the Gaussian behavior of atmospheric noise at low amplitudes is the result of many distant lightning discharges, whereas its impulsive behavior at high amplitudes is caused by strong spikes from nearby thunderstorms. A typical empirical model then approximates the probability distribution of the noise envelope by a Rayleigh distribution at low levels and a heavy-tailed distribution at high levels. In many cases, it has been observed that the heavy-tailed distribution can be assumed to follow some algebraic law x^{-n}, where n is typically between 1 and 3 [Lerner, 1961; Mertz, 1961; Ibukun, 1966].

The behavior of the $S\alpha S$ model coincides with these empirical observations, i.e., $S\alpha S$ distributions exhibit Gaussian behavior at low amplitudes and decay algebraically at the tails. But unlike the empirical models, the $S\alpha S$ model provides physical insight into the noise generation process and is not limited to particular situations. It is certainly possible that other probability distributions could be formulated exhibiting these behaviors, but the $S\alpha S$ model is preferred for several reasons. First, stable distributions share many convenient properties with the Gaussian distribution, such as the stability property, as shown earlier. In addition, they are very flexible as a modeling tool in that the characteristic exponent α allows us to represent noise with a continuous range of impulsiveness. A small value of α implies that the noise is highly impulsive, while a value of α close to 2 indicates a Gaussian type of behavior. Finally, it agrees very well with the measured data of a variety of man-made and natural noises, as demonstrated below.

Our first example is atmospheric noise, which is the predominant noise source at very low frequency (VLF) and extreme low frequency (ELF). Figure 9.5 compares the $S\alpha S$ model with experiment for typical ELF noise. The measured points for moderate-level Malta ELF noise in the bandwidth from 5 to 320 Hz are taken from Evans and Griffths [1974]. Since the ratio of bandwidth to center frequency is not small at ELF, the APD of wideband $S\alpha S$ noise given by (9.48) is used. The characteristic exponent α and the dispersion γ are selected to best fit the data. Figure 9.6 is

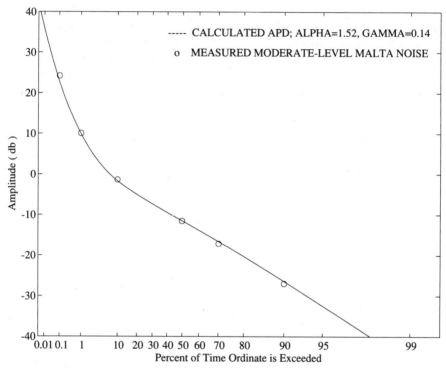

FIGURE 9.5 Comparison of a measured APD of ELF atmospheric noise with the $S\alpha S$ model.

analogous to Figure 9.5, and compares the $S\alpha S$ model with experiment for typical VLF noise. The experimental APD is replotted from Middleton [1976] and the theoretic APD is calculated from (9.51) by selecting best values of α and γ. The two figures show that the two-parameter representation of the APD by $S\alpha S$ distributions provides an excellent fit to measurements of atmospheric noise.

Similar conclusions are also applicable to other man-made and natural impulsive noise, as shown in Figures 9.7 and 9.8. Figure 9.7 is an example of primarily urban automotive ignition noise, whereas Figure 9.8 shows the APD for fluorescent lights in a mine shop [Middleton, 1976]. In both cases, the agreement between theory and experiment is very good.

Another example of $S\alpha S$ noise is the impulsive noise observed on telephone lines. These impulsive interferences are caused by several sources including lightning, switching transients, and accidental hits during maintenance work. A detailed empirical study shows that noise on several telephone lines can be adequately modeled by $S\alpha S$ distributions with characteristic exponents close to but definitely less than 2 [Stuck and Kleiner, 1974] .

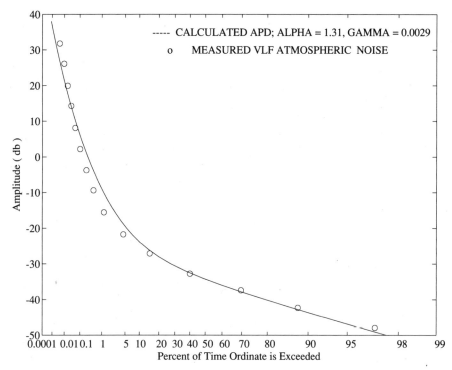

FIGURE 9.6 Comparison of a measured envelope APD of VLF atmospheric noise with the $S\alpha S$ model.

9.7 CONCLUSION

In this chapter, first-order statistics of impulsive noise processes have been developed from the filtered-impulse mechanism. Under appropriate assumptions on the spatial and temporal distributions of noise sources and the propagation conditions, we have shown that the instantaneous amplitude of the received noise obeys the symmetric stable distribution. In the case of narrowband reception, the joint distribution of the in-phase and quadrature components of the received noise is isotropic stable. The noise phase is shown to be uniformly distributed in $[0, 2\pi]$ and independent of the envelope, as in the Gaussian case. The distribution of the envelope, on the other hand, is a heavy-tailed generalization of the Rayleigh distribution. Compared with existing models, such as Middleton's statistical-physical canonical models, the symmetric stable model is much simpler and mathematically more appealing. Direct comparisons with experimental data show that this model fits closely a variety of non-Gaussian noises.

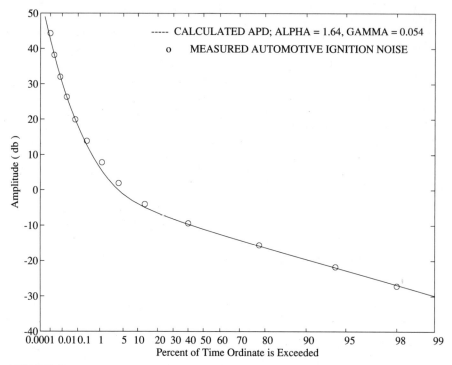

FIGURE 9.7 Comparison of a measured envelope APD of automotive ignition noise with the $S\alpha S$ model.

PROBLEMS

1. (Image processing with $S\alpha S$ distributions) For a noiseless image, lena.512, for example (you can use anonymous ftp hexam.usc.edu to obtain this image), (i) contaminate the image by adding standard Gaussian noise and then find a method to clean the noise. (ii) Contaminate the image by adding a standard $S\alpha S$ noise (let α be close to 2 for practical reasons); first use the same method in (i) as if the noise is Gaussian, then develop a method for cleaning the non-Gaussian $S\alpha S$ noise. How do the results compare? (*Hint*: If after adding noise, the gray level range is wider than $[0, 255]$, you can perform contrast manipulation at the last step.)

2. As derived in the book, the density of the $S\alpha S$ noise envelope is

$$f(a) = a \int_0^\infty s \exp(-\gamma s^\alpha) J_0(as) \, ds.$$

What is the mean of this distribution? Does it have finite second- or higher-order moments? What about fractional pth-order moments? (*Hint*: Consider both Gaussian and non-Gaussian).

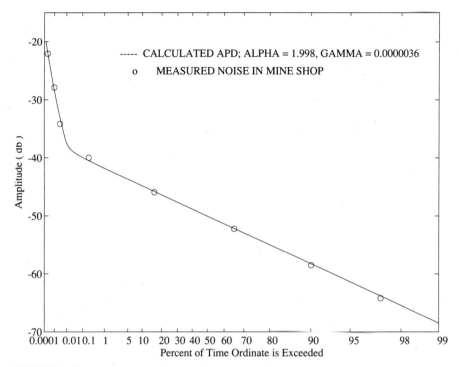

FIGURE 9.8 Comparison of a measured envelope APD of fluorescent lights in a mine shop office with the $S\alpha S$ model.

3. (i) What is the narrowband representation of Gaussian noise? What is the joint p.d.f. of the in-phase and quadrature components? Are they independent? (ii) What is the narrowband representation of a Cauchy noise? What is the joint p.d.f of the in-phase and quadrature components? Are they independent?
4. Middleton's statistical-physical canonical models can also be used to model impulsive noise environments. What is the difference between Middleton's models (Class A and Class B) and a $S\alpha S$ model?
5. The lattice structure is an efficient method of implementing the Gram-Schmidt orthogonalization method. When the input is a second-order process, the reflection coefficients can be expressed in terms of the second-order statistics of the prediction errors by minimizing their variance. When the input is a $S\alpha S$ process, the usual L^2 notion of orthogonality is inapplicable. What is the orthogonality in Banach spaces? Find the reflection coefficients when the input is a $S\alpha S$ random process.
6. (Research topic) From Shannon's *capacity theorem*, (i) what is the Gaussian channel capacity? (ii) What is the Cauchy channel capacity? (iii) What is the $S\alpha S$ channel capacity in general?

10

Signal Detection in Stable Noise

10.1 INTRODUCTION

The problem of designing optimum detectors in the presence of additive noise has a long history of investigation in the statistical signal processing literature. To simplify the implementation and analysis of the detectors, one usually assumes that the noise is Gaussian. However, as we have seen in the previous chapter, there are a number of important situations where dominant interferences are impulsive and can be characterized more accurately as $S\alpha S$ noises. It is well known that the linear detector, which is optimal under the Gaussian assumption, is no longer appropriate for an impulsive interference, because the large dynamic ranges of impulsive noise generally result in significant performance degradation, such as increased false alarm rate or error probability. In order to minimize these effects, it is necessary to incorporate some type of nonlinearity into the detector structures to reduce the impulsiveness of the noise.

The purpose of this chapter is to study the structures and performances of the nonlinear detectors. We shall pay particular attention to the so-called locally optimum detector for the detection of weak signals under the assumption that the probability distribution of the corrupting noise is symmetric stable. Because of the complications in the implementation of the locally optimum detector, one often uses more easily implementable suboptimum detectors with nonlinearities such as clippers, hole puncher, and hard limiters. In this chapter, we will also investigate the performances of selected suboptimum nonlinear detectors in the presence of stable noise and compare them with that of the locally optimum receiver.

10.2 LOCALLY OPTIMUM AND SUBOPTIMUM DETECTORS

Consider the problem of detecting the presence of a known signal $s(t)$ over the interval $[0, T]$ in the presence of additive noise. It may be mathematically formulated as a problem of testing an alternative hypothesis H_1 versus a null hypothesis H_0,

10.2 LOCALLY OPTIMUM AND SUBOPTIMUM DETECTORS

where

$$H_1 : x(t) = \theta s(t) + n(t)$$
$$H_0 : x(t) = n(t) \qquad \theta > 0, \quad 0 \le t \le T. \qquad (10.1)$$

Under H_1, the received waveform $x(t)$ consists of noise plus known signal $s(t)$ with a deterministic but unknown magnitude θ, and under H_0 the observation consists of noise only. To facilitate the design and analysis of detectors, we shall follow the standard procedure of replacing all continuous waveforms with vectors of N samples to obtain the following expressions for the hypotheses

$$H_1 : \mathbf{x} = \theta \mathbf{s} + \mathbf{n}, \qquad \theta > 0$$
$$H_0 : \mathbf{x} = \mathbf{n} \qquad (10.2)$$

where all the vectors have N samples. To further simplify the problem, we shall assume that the sampling period is large enough so that the noise vector $\mathbf{n}$ contains i.i.d. samples, each having density function $f_n(x)$.

The task of a detector is to choose between H_1 and H_0 based solely on the observation $\mathbf{x}$ and without knowledge of the unknown amplitude θ. In doing so, it can make two types of errors. A type I error is made if the detector accepts H_1 when in fact H_0 is correct. The probability of this type of error is called the *false-alarm probability*, denoted by p_f. A type II error is accepting H_0 when H_1 is correct. The probability of this error is $1 - p_d(\theta)$, where $p_d(\theta)$ is the *detection probability* of accepting H_1 when the signal is indeed present with amplitude θ. Note that the false-alarm probability p_f is independent of the signal amplitude θ while the detection probability $p_d(\theta)$ is generally a function of θ. The false-alarm and detection probabilities may be conveniently combined as the *power function* of the detector as follows: for $\theta \ge 0$

$$p(\theta) = \begin{cases} p_d(\theta) & \text{if } \theta > 0 \\ p_f & \text{if } \theta = 0. \end{cases} \qquad (10.3)$$

Hence, the performance of the detector is completely specified by its power function.

An ideal detector has a power function that is zero for $\theta = 0$ and is unity when $\theta > 0$. Unfortunately, such an ideal detector does not exist since these are two conflicting requirements. Instead, optimum detectors are usually designed by maximizing the power function $p(\theta)$ for $\theta > 0$ (i.e., the detection probability) for a given false-alarm probability $p_f = p(0)$. This design philosophy is used, for example, in radar system design where a type II error is much more costly than a type I error. Thus, for any two detectors for the hypotheses in (10.2) with power functions $p_1(\theta)$ and $p_2(\theta)$, respectively, and the same false-alarm probability $p_f = p_1(0) = p_2(0)$, the first detector is said to be *better*, or *more powerful*, than the second one for the given false-alarm probability p_f if

$$p_1(\theta) \ge p_2(\theta), \quad \forall \theta > 0.$$

A detector is said to be *uniformly most powerful* (UMP) if it is more powerful than any other detectors for the given false-alarm probability.

Although UMP detectors are very desirable, they rarely exist. For any two detectors with power functions $p_1(\theta)$ and $p_2(\theta)$ and the same false-alarm probability, the most common scenario is that $p_1(\theta) > p_2(\theta)$ for some values of the signal strength θ and $p_1(\theta) < p_2(\theta)$ for others. One of the exceptional cases where a UMP detector does exist is when the value of $\theta = \theta_0$ is known. In this case, by the Neyman-Pearson lemma [Ferguson, 1967], the UMP detector for the hypotheses in (10.2) is given by the following log likelihood ratio test

$$T_{NP}(\mathbf{x}) = \sum_{i=1}^{N} \ln \frac{f_n(x_i - \theta_0 s_i)}{f_n(x_i)} \underset{H_0}{\overset{H_1}{\gtrless}} t \qquad (10.4)$$

where the threshold t is determined by the given false-alarm probability. When the noise is Gaussian, the UMP detector also exists and is the well-known linear correlator (matched filter), namely,

$$T_{LC}(\mathbf{x}) = \sum_{i=1}^{N} s_i x_i > t \qquad (10.5)$$

where the threshold t is again determined by the given false-alarm probability. Note that the above decision rule is independent of the value of θ.

When UMP detectors are unavailable, more realistic criteria have to be used in designing optimum detectors. One such criterion is the so-called *local optimality*. It is based on the observation that in many applications, such as spread spectrum communications, the signal is very weak, i.e., θ is a very small positive number. In such situations, the values of the power function $p(\theta)$ of a detector for small θ are much more important than its values for large θ. Hence, instead of maximizing $p(\theta)$ for all $\theta > 0$, one needs only to maximize $p(\theta)$ for $0 < \theta \leq \theta_0$ for some $\theta_0 \approx 0$. To be more specific, let $p_1(\theta)$ and $p_2(\theta)$ be the power functions of two detectors with the same false-alarm probability $p_f = p_1(0) = p_2(0)$. The first detector is said to be *locally more powerful* than the second one if there exists a $\theta_0 > 0$ such that

$$p_1(\theta) \geq p_2(\theta), \qquad \forall 0 < \theta \leq \theta_0.$$

Naturally, a *locally optimum (most powerful)* (LO) detector is the one that is locally more powerful than any other detectors.

Unlike UMP detectors, locally optimum detectors always exist. To see this, one notes that a detector is locally optimum if the first-order derivative of its power function $p(\theta)$ at $\theta = 0$ is maximal for a fixed false-alarm probability $p(0)$. Following a simple generalization of the Neyman-Pearson lemma [Ferguson, 1967; Casella and Berger, 1990], one can show that the locally optimum detector has the

10.2 LOCALLY OPTIMUM AND SUBOPTIMUM DETECTORS

following structure

$$T(\mathbf{x}) = \sum_{i=1}^{N} -\frac{f_n'(x_i)}{f_n(x_i)} s_i \underset{H_0}{\overset{H_1}{\gtrless}} t \quad (10.6)$$

where $f_n'(x)$ is first-order derivative of the noise density $f_n(x)$ and t is a threshold determined by the pre-specified false-alarm probability.

The optimality of LO detectors when the signal is weak can be seen more clearly through their relations with the Neyman-Pearson likelihood ratio detector. Recall that for each known value of θ, the Neyman-Pearson lemma suggests that the UMP detector is given by (10.4). Now suppose that θ is very small, so one uses a Taylor expansion to approximate the log-likelihood, i.e.,

$$T_{NP}(\mathbf{x}) \approx \left(\sum_{i=1}^{N} -\frac{f_n'(x_i)}{f_n(x_i)} s_i \right) \theta.$$

Since the constant θ can be absorbed into the threshold, it is clear that the locally optimum and UMP detectors coincide. In other words, the LO detector is uniformly most powerful when the signal strength is small. A rigorous proof of this result is given in [Capon, 1961]. In the case of Gaussian noise, the locally optimum and UMP detectors coincide for all values of the signal strength.

The LO detector has the form of a general class of nonlinear detectors shown in Figure 10.1. It is only a slight modification of the optimum linear detector for Gaussian noise, consisting of a linear structure proceeded by a zero-memory non-linearity (ZMNL). Yet it is also general enough to include most of practical linear and nonlinear detectors. With an identity nonlinearity, one of course obtains the

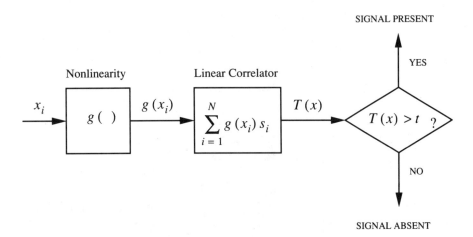

FIGURE 10.1 Typical nonlinear detector structure.

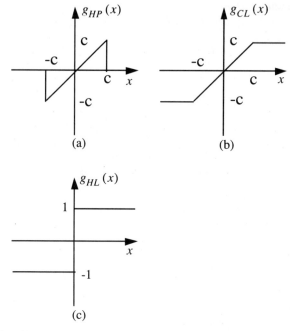

FIGURE 10.2 Simple nonlinearities: (a) hole puncher, (b) clipper, and (c) hard limiter.

familiar matched filter that is optimal for Gaussian noise. For the locally optimum detector, the nonlinearity function can be determined from the noise probability density by

$$g(x) = -f_n'(x)/f_n(x). \tag{10.7}$$

Commonly used nonlinearity functions include the hole puncher, clipper, and hard-limiter, shown in Figure 10.2. The hard-limiter, for example, is the optimum nonlinearity for the LO detector when the noise has a double-exponential density of the form

$$f_n(x) = \frac{\lambda}{2}\exp(-\lambda|x|), \quad \lambda > 0.$$

10.3 PERFORMANCE OF LOCALLY OPTIMUM AND SUBOPTIMUM DETECTORS FOR STABLE NOISE

Let us now assume that the noise is $S\alpha S$ with dispersion γ, and apply the LO detector to the hypotheses problem specified in (10.2). In this case, the optimum nonlinearity is given by

$$g_{LO}(x) = -\frac{f_{\alpha,\gamma}'(x)}{f_{\alpha,\gamma}(x)} \tag{10.8}$$

10.3 PERFORMANCE OF OPTIMUM AND SUBOPTIMUM DETECTORS

where $f_{\alpha,\gamma}(x)$ is the $S\alpha S$ density in (2.9). Although there is no explicit expression for the above nonlinearity, some of its asymptotic properties can be readily obtained. For example, since the stable density function is approximately Gaussian around the origin, the nonlinearity $g_{LO}(x)$ for the LO detector is roughly a linear function for small inputs. And the closer the characteristic exponent α is to 2, the wider the linear range. When $\alpha = 2$, i.e., when the noise is Gaussian, $g_{LO}(x)$ is a linear function as expected. For large inputs, since $f_{\alpha,\gamma}(x) \approx C|x|^{-\alpha-1}$ as $|x| \to \pm\infty$ [Shao and Nikias, 1993a], one has

$$g_{LO}(x) \to \frac{\alpha+1}{x}, \quad \text{as } x \to \pm\infty.$$

Namely, the output of the locally optimum nonlinearity is asymptotically inversely proportional to the input and thus suppresses large inputs. All these characteristics are clearly shown in Figure 10.3, which plots the optimum nonlinearity for various values of α and γ.

Due to the lack of closed-form expression for the optimum nonlinearity $g_{LO}(x)$, the implementation of the LO detector in the case of stable noise is complicated. In the case where the simplicity in the implementation of the detector is of primary concern, one may have to replace the optimum nonlinearity with suboptimum nonlinearities such as those shown in Figure 10.2. Because of the characteristics of the locally optimum nonlinearity, the hole puncher provides a good approximation, especially when the input is small. The range of the hole puncher may be determined by cutting off the decreasing portion of the locally optimum nonlinearity curve. An example is shown in Figure 10.4 for $\alpha = 1.9, \gamma = 1$, where, as expected, the locally optimum nonlinearity and the hole puncher agree well around the origin. Of course, one may still choose to use the linear correlator. In any case it is of considerable interest to compare the performances of these nonlinear detectors.

There are generally two approaches to compare the performances of detectors, i.e., asymptotic approach and finite sample approach. In the finite sample approach, the number of observation N is assumed to be finite and fixed, while the asymptotic approach assumes $N \to \infty$. In both cases, the performances of detectors are compared by their detection probability for a given false-alarm probability and signal strength. Under the asymptotic assumption, one may invoke the central limit theorem to get closed-form results. These results, however, are valid only for the case when the sample size is large. The finite sample approach is more realistic since only finite number of samples are available in practice. In this section we shall take the finite-sample approach and compare the performances of the locally optimum (LO), linear correlator (LC), and hole puncher (HP) detector.

Recall that the test statistics of the LO, LC, and HP detectors are

$$T_{LO}(\mathbf{x}) = \sum_{i=1}^{N} g_{LO}(x_i)s_i \tag{10.9}$$

$$T_{LC}(\mathbf{x}) = \sum_{i=1}^{N} x_i s_i, \tag{10.10}$$

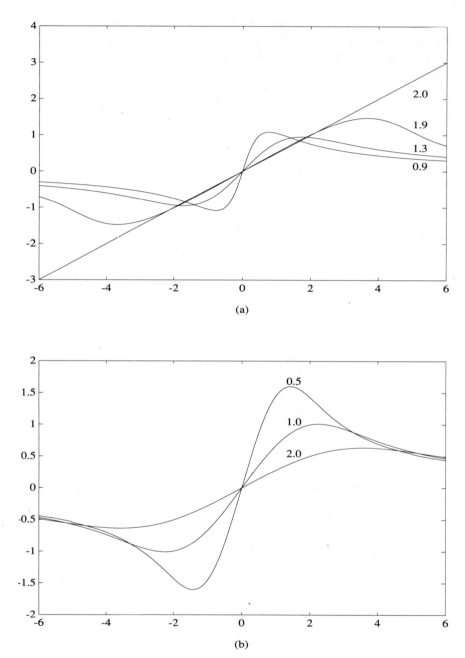

FIGURE 10.3 Locally optimum nonlinearity for the stable noise: (a) $\gamma = 1$ and various values of α, and (b) $\alpha = 1.5$ and various values of γ.

10.3 PERFORMANCE OF OPTIMUM AND SUBOPTIMUM DETECTORS

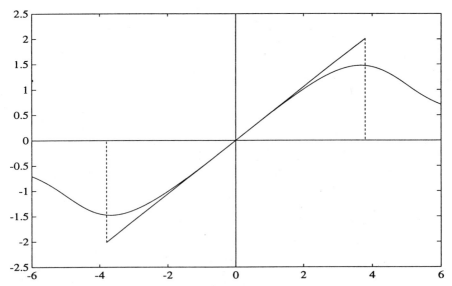

FIGURE 10.4 Locally optimum nonlinearity for the stable noise with $\alpha = 1.9, \gamma = 1$ and the corresponding approximating hole puncher.

and

$$T_{HP}(\mathbf{x}) = \sum_{i=1}^{N} g_{HP}(x_i) s_i \qquad (10.11)$$

respectively, where g_{LO} is given by (10.8) and

$$g_{HP}(x) = \begin{cases} x & \text{if } |x| < c \\ 0 & \text{otherwise.} \end{cases} \qquad (10.12)$$

Since the closed-form evaluation of the conditional probability density function of the statistics is rather intractable under both hypotheses, the performances (detection probability $p_d(\theta)$ and false-alarm probability p_f) of the detectors have been carried out by Monte Carlo simulations.

In the simulations, the number of signal components N is fixed to be 20, and the known signal components are assumed to be $s_i = 1, i = 1, \ldots, N$. False-alarm probability is set to 10^{-2} so the threshold of each detector can be found through a Monte Carlo simulation. After obtaining the thresholds, the detection probability for each signal strength θ is again found through a Monte Carlo simulation.

The performances of the LO, LC, and HP detectors are shown in Figures 10.5 and 10.6 for $\alpha = 1.9$ and $\alpha = 1.3$. From these two figures, we observe that the performance of the LO detector is always better than that of the HP and LC detectors. The smaller α, the better the LO detector compared with the other two

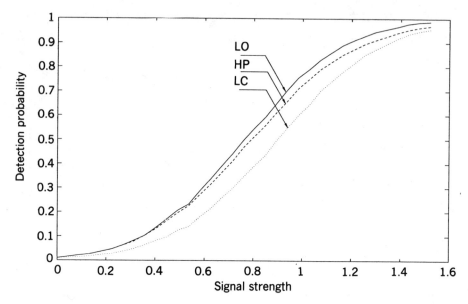

FIGURE 10.5 Detection probability of detectors in $S\alpha S$ noise with $\alpha = 1.9, \gamma = 1.0$ and false alarm probability 10^{-2}.

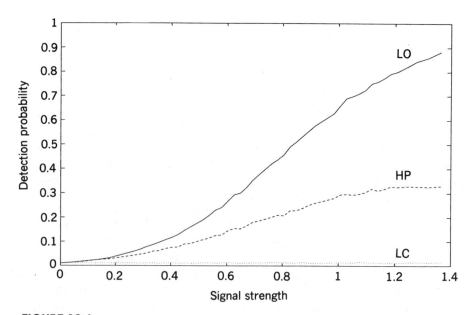

FIGURE 10.6 Detection probability of detectors in $S\alpha S$ noise with $\alpha = 1.3, \gamma = 1.0$ and false alarm probability 10^{-2}.

detectors. It should be borne in mind that the performance gain is obtained at the expense of computational complexity. The HP detector also uniformly outperforms the LC detector. In fact, when α is small, the LC detector is almost useless. When α is close to 2, the performance of the HP detector is close to that of the LO detector, especially for weak signals, despite its computational simplicity. This is expected since in this case the nonlinearity of the HP detector approximates the LO nonlinearity pretty well.

10.4 ASYMPTOTIC PROBABILITY OF ERROR

In this section, we examine the performance of coherent receivers in independent $S\alpha S$ noise [Tsihrintzis and Nikias, 1995b]. For simplicity, we restrict the presentation to binary signaling; however, the generalization to arbitrary M-ary signaling and to incoherent reception is straightforward and the results presented here will also hold for those cases.

Our mathematical model is the following hypothesis testing problem:

$$H_0 : x(k) = s_0(k) + n_\alpha(k), \quad k = 1, 2, \ldots, N$$

$$H_1 : x(k) = s_1(k) + n_\alpha(k), \quad k = 1, 2, \ldots, N,$$

where $s_i(\cdot)$, $i = 0, 1$, is one of two possible transmitted signals and $n_\alpha(\cdot)$ is a realization of a sequence of N independent, identically distributed $S\alpha S$ random variables of characteristic exponent α ($0 < \alpha \leq 2$) and dispersion γ. The receiver needs to make a decision on which hypothesis is true (i.e., which signal $s_i(\cdot)$ was sent) on the basis of the observed data $x(\cdot)$. We will assume that $s_1(k) = -s_0(k)$, $k = 1, 2, \ldots, N$ (antipodal signaling), as is for example the case of a BPSK communications system. The following analysis is, however, valid for arbitrary signaling waveforms and is, thus, applicable to all the communications systems in use.

10.4.1 Optimum Receiver

To decide between the two hypotheses H_0 and H_1, the optimum (in the maximum likelihood (ML) sense) receiver computes the test statistic

$$\begin{aligned}\Lambda &= \log\left\{\frac{\prod_{k=1}^N f_\alpha[x(k) - s_0(k)]}{\prod_{k=1}^N f_\alpha[x(k) - s_1(k)]}\right\} \\ &= \sum_{k=1}^N \log\left\{\frac{f_\alpha[x(k) - s_0(k)]}{f_\alpha[x(k) - s_1(k)]}\right\}\end{aligned} \quad (10.13)$$

and compares it to a preset threshold η. When $\Lambda \geq \eta$, the receiver decides that $s_0(\cdot)$ was sent; otherwise that $s_1(\cdot)$ was sent. We are going to assume that $\eta = 0$, a threshold setting which minimizes the probability of error. The following analysis is valid, however, for arbitrary threshold η.

Clearly, the test statistic Λ is a random variable with finite mean and variance, even when the noise process n_α is a $S\alpha S$ process with infinite variance [Rappaport and Kurz, 1966].[1] The mean of the test statistic Λ, assuming $s_0(\cdot)$ was sent, can then be computed as

$$\mu_0 = \mathbf{E}\{\Lambda | s_0(\cdot) \text{ sent}\}$$

$$= \sum_{k=1}^{N} \int_{-\infty}^{\infty} f_\alpha(\xi - s_0(k)) \log \left\{ \frac{f_\alpha[\xi - s_0(k)]}{f_\alpha[\xi - s_1(k)]} \right\} d\xi$$

$$= \sum_{k=1}^{N} \int_{-\infty}^{\infty} f_\alpha(\xi) \log \left\{ \frac{f_\alpha(\xi)}{f_\alpha[\xi + 2s_0(k)]} \right\} d\xi$$

$$= -\mathbf{E}\{\Lambda | s_1(\cdot) \text{ sent}\} = -\mu_1 \qquad (10.14)$$

and its finite variance as

$$\sigma^2 = \text{var}\{\Lambda | s_0(\cdot) \text{ or } s_1(\cdot) \text{ sent}\}$$

$$= \sum_{k=1}^{N} \int_{-\infty}^{\infty} f_\alpha(\xi) \log^2 \left\{ \frac{f_\alpha(\xi)}{f_\alpha[\xi + 2s_0(k)]} \right\} d\xi$$

$$- \sum_{k=1}^{N} \left[\int_{-\infty}^{\infty} f_\alpha(\xi) \log \left\{ \frac{f_\alpha(\xi)}{f_\alpha[\xi + 2s_0(k)]} \right\} d\xi \right]^2. \qquad (10.15)$$

The test statistic Λ being a superposition of N independent random variables satisfying the assumptions of the Central Limit Theorem, its asymptotic (for large N) distribution is Gaussian with mean and variance given by (10.14) and (10.15), respectively. Assuming that the signals $s_0(\cdot)$ and $s_1(\cdot)$ are sent with equal probability (equal to $\frac{1}{2}$), the average probability of an erroneous decision will be

$$P_e = \Pr\{\Lambda < 0 | s_0(\cdot) \text{ sent}\} = \frac{1}{2} \text{erfc}\left(\frac{\mu_0}{\sqrt{2\sigma^2}} \right), \qquad (10.16)$$

where erfc $(\cdot)$ is the complimentary error function: $\text{erfc}(x) = \frac{2}{\sqrt{\pi}} \int_x^\infty e^{-\xi^2} d\xi$.

10.4.2 Linear (Gaussian) Receiver

The linear receiver is derived by assuming that the characteristic exponent of the noise p.d.f. $f_\alpha(\cdot)$ in the test statistic Λ of (10.13) equals $\alpha = 2$. Thus, the lin-

---
[1] One way to understand this assertion is to consider the asymptotic behavior of the integrands in (10.14) and (10.15) as $\xi \to \infty$.

ear receiver is the optimum receiver in the ML sense when the interfering noise is Gaussian. We will now examine its performance when the incoming signal is corrupted by $S\alpha S$ noise of characteristic exponent α not necessarily equal to 2 and compare it to the performance of the corresponding optimum receiver of the previous subsection.

After setting $\alpha = 2$ and carrying out the algebra in (10.13), we end up with the expression

$$\Lambda_l = \frac{1}{\gamma} \sum_{k=1}^{N} x(k) s_0(k) \tag{10.17}$$

for the test statistic of the linear receiver. In (10.17) above, γ is the dispersion in the interfering $S\alpha S$ noise. Since the test statistic Λ_l is a linear superposition of N independent $S\alpha S$ random variables, it is itself a $S\alpha S$ random variable. Its location parameter, assuming $s_0(\cdot)$ was sent, is

$$\delta_0 = \frac{1}{\gamma} \sum_{k=1}^{N} s_0^2(k) = -\delta_1 \tag{10.18}$$

and its dispersion

$$\gamma_l = \gamma^{1-\alpha} \sum_{k=1}^{N} |s_0(k)|^\alpha. \tag{10.19}$$

Thus, assuming again equiprobable signals $s_0(\cdot)$ and $s_1(\cdot)$, the average probability of erroneous decision at the linear receiver becomes

$$P_e^l = \Pr\{\Lambda_l < 0 | s_0(\cdot) \text{ sent}\} = \int_{-\infty}^{0} f_\alpha(\gamma_l, \delta_0; \xi) \, d\xi, \tag{10.20}$$

where $f_\alpha(\gamma_l, \delta_0; \cdot)$ is the $S\alpha S$ p.d.f. with location parameter δ_0 and dispersion γ_l.

10.4.3 Limiter Plus Integrator

Let us consider a limiting device with input-output characteristic

$$g(\xi) = \begin{cases} -\kappa, & \text{if } \xi < -\kappa \\ \xi, & \text{if } |\xi| < \kappa \\ \kappa, & \text{if } \xi > \kappa. \end{cases}$$

Let the input to this limiter be a random variable x of p.d.f. $f_{in}(\cdot)$ and cumulative distribution function $F_{in}(\cdot)$. At the output of the limiter, a random variable y will then be observed with p.d.f. [Papoulis, 1991]

$$f_{out}(\xi) = [1 - F_{in}(\kappa)]\delta(\xi - \kappa) + F_{in}(-\kappa)\delta(\xi + \kappa) + I_{(-\kappa,\kappa)}(\xi)f_{in}(\xi), \quad (10.21)$$

where $I_{(-\kappa,\kappa)}(\xi) = 1$, when $|\xi| < \kappa$, and zero, otherwise.

In our case, let $y(\cdot)$ be the output of the limiter when the datum $x(\cdot)$ is observed, i.e., $y(k) = g[x(k)], k = 1, 2, \ldots, N$. Then, $y(\cdot)$ is a sequence of independent, identically distributed random variables, each having a p.d.f. given by (10.21) with $f_{in}(\cdot)$ ($F_{in}(\cdot)$) being a SαS p.d.f. $f_\alpha(\gamma, \delta_i; \cdot)$ (cumulative distribution function $F_\alpha(\gamma, \delta_i; \cdot)$) of dispersion γ and location parameter δ_i. Here, $i = 0$ or 1, depending on whether $s_0(\cdot)$ or $s_1(\cdot)$ was sent. The receiver bases its decision on the test statistic

$$\Lambda_{LI} = \sum_{k=1}^{N} y(k)s_0(k) \quad (10.22)$$

and decides that $s_0(\cdot)$ was sent if $\Lambda_{LI} \geq 0$; otherwise $s_1(\cdot)$ was sent. Assuming that $s_0(\cdot)$ was sent, the test statistic has mean

$$\begin{aligned}
\mu_{0,LI} &= \mathbf{E}\{\Lambda_{LI}|s_0(\cdot) \text{ sent}\} \\
&= \sum_{k=1}^{N} s_0(k)\mathbf{E}\{y(k)|s_0(\cdot) \text{ sent}\} \\
&= -\mathbf{E}\{\Lambda_{LI}|s_1(\cdot) \text{ sent}\} \\
&= -\mu_{1,LI}
\end{aligned} \quad (10.23)$$

and finite variance

$$\begin{aligned}
\sigma_{LI}^2 &= \text{var}\{\Lambda_{LI}|s_0(\cdot) \text{ or } s_1(\cdot) \text{ sent}\} \\
&= \sum_{k=1}^{N} |s_0(k)|^2 \text{ var}\{y(k)|s_0(\cdot) \text{ or } s_1(\cdot) \text{ sent}\}.
\end{aligned} \quad (10.24)$$

The mean and variance, $\mathbf{E}\{y(k)|s_0(\cdot) \text{ sent}\}$ and $\text{var}\{y(k)|s_0(\cdot) \text{ or } s_1(\cdot) \text{ sent}\}$, can be directly computed using (10.21) as

$$\begin{aligned}
\mathbf{E}\{y(k)|s_0(\cdot) \text{ sent}\} &= \int_{-\infty}^{\infty} \xi f_{out}(\xi|s_0(\cdot) \text{ sent}) \, d\xi \\
&= \kappa - \int_{-\kappa}^{\kappa} F_\alpha(\gamma, s_0(k); \xi) \, d\xi
\end{aligned} \quad (10.25)$$

$$\text{var}\{y(k)|s_0(\cdot) \text{ or } s_1(\cdot) \text{ sent}\} = \int_{-\infty}^{\infty} \xi^2 f_{out}(\xi|s_0(\cdot) \text{ sent}) d\xi$$

$$- \mathbf{E}^2\{y(k)|s_0(\cdot) \text{ sent}\}$$

$$= \kappa^2 - 2 \int_{-\kappa}^{\kappa} \xi F_\alpha(\gamma, s_0(k); \xi) d\xi$$

$$- \mathbf{E}^2\{y(k)|s_0(\cdot) \text{ sent}\}. \quad (10.26)$$

Substituting (10.25) and (10.26) into (10.23) and (10.24), we get

$$\mu_{0,LI} = \sum_{k=1}^{N} s_0(k) \left[\kappa - \int_{-\kappa}^{\kappa} F_\alpha(\gamma, s_0(k); \xi) d\xi \right] \quad (10.27)$$

and

$$\sigma_{LI}^2 = \sum_{k=1}^{N} |s_0(k)|^2 \left[\kappa^2 - 2 \int_{-\kappa}^{\kappa} \xi F_\alpha(\gamma, s_0(k); \xi) d\xi \right] - \mu_{0,LI}^2. \quad (10.28)$$

Assuming again equiprobable signals $s_0(\cdot)$ and $s_1(\cdot)$, the asymptotic (for large N) probability of error for this receiver will be

$$P_e^{LI} = \frac{1}{2} \text{erfc} \left(\frac{\mu_{0,LI}}{\sqrt{2\sigma_{LI}^2}} \right). \quad (10.29)$$

10.4.4 Cauchy Receiver

By Cauchy receiver we mean the receiver that employs as a test the statistic Λ_C derived from (10.13) under the assumption that $\alpha = 1$ (Cauchy noise). Given this definition and the Cauchy distribution, we get

$$\Lambda_C = \sum_{k=1}^{N} \log \left\{ \frac{f_1[x(k) - s_0(k)]}{f_1[x(k) - s_1(k)]} \right\} = \sum_{k=1}^{N} \log \left\{ \frac{\gamma^2 + [x(k) + s_1(k)]^2}{\gamma^2 + [x(k) - s_0(k)]^2} \right\}. \quad (10.30)$$

The performance of the Cauchy receiver can be analyzed following steps similar to those followed in the analysis of the optimum receiver. In fact, the same approach can be taken to analyze the performance of a general SαS receiver constructed with a characteristic exponent mismatched to the actual characteristic exponent of the interfering noise. Following steps similar to those of the previous section, we end up with the expression for the probability of error of the Cauchy receiver:

$$P_e^C = \frac{1}{2}\text{erfc}\left(\frac{\mu_{0,C}}{\sqrt{2\sigma_c^2}}\right), \tag{10.31}$$

where

$$\mu_{0,C} = \sum_{k=1}^{N} \int_{-\infty}^{\infty} f_\alpha(\xi) \log\left\{\frac{f_1(\xi)}{f_1[\xi + 2s_0(k)]}\right\} d\xi \tag{10.32}$$

and

$$\sigma_C^2 = \sum_{k=1}^{N} \int_{-\infty}^{\infty} f_\alpha(\xi) \log^2\left\{\frac{f_1(\xi)}{f_1[\xi + 2s_0(k)]}\right\} d\xi$$
$$- \sum_{k=1}^{N} \left[\int_{-\infty}^{\infty} f_\alpha(\xi) \log\left\{\frac{f_1(\xi)}{f_1[\xi + 2s_0(k)]}\right\} d\xi\right]^2. \tag{10.33}$$

10.4.5 A Note on the Exact Computation of the Probability of Error

With the exception of the linear receiver, the expressions for the probability of error that we derived in previous section are only asymptotically valid, i.e., they hold true only when the length N of the data sequence is large enough for the true distribution of the several test statistics to be well approximated by a Gaussian distribution. Thus, a minimum number N of data samples may be needed for the expressions in (10.16), (10.29), and (10.31) to be meaningful (see, for example, Papoulis [1991] for an illustration of the convergence rate of the central limit theorem). In general, we expect an error in the predicted values for the probabilities of error of the different receivers to arise from the sensitivity of these probabilities to the far tails of the true p.d.f. of the corresponding test statistics, which are only poorly approximated by the Gaussian p.d.f. [Helstrom and Ritcey, 1984]. A better estimate of the probabilities of error of the several receivers can be obtained either by performing extensive Monte Carlo simulations or by the characteristic function based method that we describe next (see also Bird [1982] for similar computations).

Let t be any test statistic of the set $\{\Lambda, \Lambda_I, \Lambda_{LI}, \Lambda_C\}$. Since t is a nonlinear transformation of the random vector $x(\cdot)$ of independent components, its corresponding characteristic function will then be

$$\mathbf{E}\{e^{i\omega t}|s_0(\cdot) \text{ sent}\} = \int_{-\infty}^{\infty} \cdots \int_{-\infty}^{\infty} e^{i\omega t(\xi_1,\ldots,\xi_N)}$$
$$\left[\prod_{k=1}^{N} f_\alpha(\gamma, s_0(k); \xi_k)\right] d\xi_1 \ldots d\xi_k \equiv T(\omega). \tag{10.34}$$

10.4 ASYMPTOTIC PROBABILITY OF ERROR

In general, the characteristic function $T(\cdot)$ can be computed numerically for a large number of values of the parameter ω and stored. After this task has been completed, the probability of error, when one uses the statistic t, is computed, again via numerical integration, as [Kendall and Stuart, 1958]

$$P_e^t = \Pr\{t < 0 | s_0(\cdot) \text{ sent}\} = \frac{1}{2} - \frac{1}{\pi} \int_0^\infty \frac{d\omega}{\omega} \Im\{T(\omega)\}, \quad (10.35)$$

where $\Im$ denotes the imaginary part. This procedure is extremely computationally intensive, requiring extensive numerical integrations, and we have chosen not to follow it. Instead, we performed comprehensive Monte Carlo simulations of all the receivers in various $S\alpha S$ noise environments and we discuss our findings in the following section.

10.4.6 Performance Comparisons

We have evaluated the expressions in (10.16), (10.20), (10.29), and (10.31), for the probability of erroneous decision for $s_0(\cdot)$ a square pulse of unit height ($s_0(k) = 1$, $k = 1, 2, \ldots, N$) and for different values of the characteristic exponent α and the dispersion γ in the p.d.f. of the additive $S\alpha S$ noise. In particular, we examined the cases of $\alpha = 0.5, 1, 1.5, 1.99$, and 2 and of $\gamma = 1, 2, \ldots, 10$. We have assumed that the decision is based on $N = 10$ samples of the incoming signal plus noise, a value that is high enough for validity of the asymptotic probabilities of error of the different receivers. The choice of an appropriate threshold for the limiter is not entirely straightforward. If this threshold is set to a very high value, then the limiter may not clip enough noise, especially in situations of very impulsive noise of high level, leading to a poor performance. On the other hand, if this threshold is set to a very low value, the limiter may overdistort the signal, especially in situations of high signal-to-noise ratio, leading again to poor performance. Thus, an optimization procedure has to be followed, in which the threshold that minimizes the average probability of error of the receiver is found. The value of this "optimum" threshold will depend significantly on the impulsiveness of the additive noise, the signal-to-noise ratio, and the transmitted pulse shape. In our simulations, the threshold of the limiter was set to $\kappa = 30$ (high value).[2] In Figures 10.7 and 10.8, we show the performance of the optimum, the linear, the limiter plus integrator, and the Cauchy receiver for the above values of α and γ. Clearly, the optimum receiver outperforms the other three types of receivers in providing a minimum probability of error for all different values of the parameters α and γ of the interfering noise. This result is, of course, compatible with the well-known fact [Van Trees, 1968] that the ML receiver also minimizes the probability of erroneous decision. The limiter plus integrator receiver provides an improvement over the linear receiver; however, its performance remains well below that of the optimum receiver. A very interesting and significant result concerning the Cauchy receiver is revealed from a

---

[2] A similar limited study of the selection of the limiter threshold, based on the processing of real data, can be found in Bouvet [1989].

comparative study of Figures 10.7 and 10.8. Clearly, *the performance of the Cauchy receiver remains only slightly below the performance of the corresponding optimum receiver and well above the performance of the limiter plus integrator receiver, even when operating in noise of arbitrary characteristic exponent* α. This fact makes the Cauchy receiver very robust for operation in environments of stable noise of unknown and/or varying characteristic exponent and is an example of the more general fact of the robustness of maximum likelihood signal processing algorithms to modeling errors [Tsihrintzis and Nikias, 1995a].

To verify the accuracy and the validity of the asymptotic expressions, we ran extensive Monte Carlo simulations of all receivers for all the previous values of the parameters α and γ of the interfering noise. In particular, random sequences of transmitted symbols of total length of 50000 were generated, with each symbol in the sequence being with the same probability ($\frac{1}{2}$) equal to one of two possibilities. For each symbol transmitted, the corresponding pulse $s_0(\cdot)$ or $s_1(\cdot)$ was generated and independent SαS stable noise, generated with the techniques of Sec. 2.2, was added to it. The output of the several receivers was then computed and, finally, the number of erroneous decisions for each receiver were counted to obtain an estimate of the performance of the receivers. Because long (50000 symbols) pseudo-random symbol sequences were employed, the estimates of the average probabilities of error are quite reliable. In fact, a comparison of the theoretical probability of error of the linear receiver, the corresponding expression of which (10.20) is exact, with the estimated probability of error obtained via Monte Carlo simulation demonstrates excellent agreement. In Figures 10.9–10.11, we compare the performance of the optimum, the linear, the limiter plus integrator (with two different threshold settings: (i) $\kappa = 30$ and (ii) $\kappa = 1$), and the Cauchy receivers operating in independent SαS noise with characteristic exponent α and dispersion γ as in the previous paragraph. A comparative study of Figures 10.7 and 10.8 and the corresponding Figures 10.9–10.11 lead to the following conclusions: (i) The theoretical performance of the linear receiver is exactly verified by Monte Carlo simulation. This, of course, was expected since the theoretical expression in (10.15) is exact and not asymptotic. (ii) The theoretical expressions for the other (nonlinear) receivers examined here are generally in agreement with the experimental ones at low noise dispersions, but diverge slightly at higher dispersions. This is, probably, due to the earlier mentioned fact that the exact probabilities of erroneous decision depend on the tails of the true distribution of the test statistics for which the (asymptotic) Gaussian approximation is not valid. As the noise dispersion increases, the tails of the test statistics become heavier and the Gaussian approximation worse.

To facilitate the comparison of the relative performance of the various receivers, we give in Figures 10.12–10.14 plots of the average probability of error of all receivers in SαS noise environments of various values of the parameters α and γ. These figures clearly demonstrate the robustness of the Cauchy receiver in stable noise of unknown/varying characteristic exponent that was mentioned in the previous paragraph. We proceed to examine the robustness of the Cauchy receiver to additional mismatches in the dispersion of the stable noise, as well as

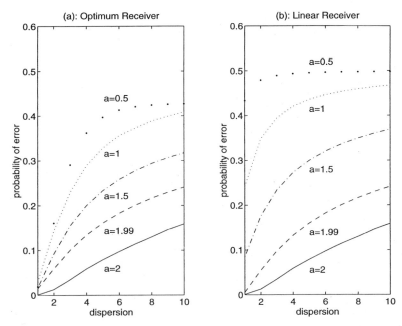

FIGURE 10.7 The asymptotic performance of the (a) optimum and (b) linear (Gaussian) receivers in the presence of $S\alpha S$ noise.

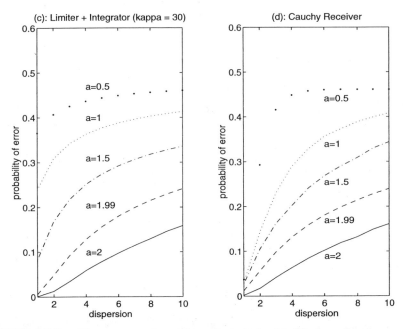

FIGURE 10.8 The asymptotic performance of the (c) "limiter plus integrator" and (d) Cauchy receivers in the presence of $S\alpha S$ noise.

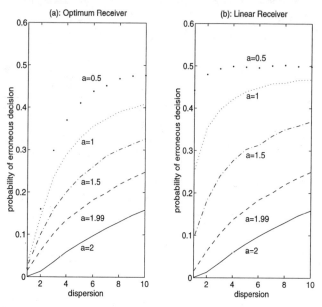

FIGURE 10.9 The performance of the (a) optimum and (b) linear (Gaussian) receivers in the presence of $S\alpha S$ noise evaluated with Monte Carlo simulation experiments.

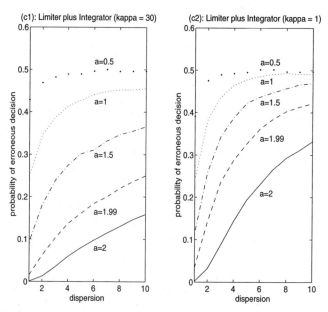

FIGURE 10.10 The performance of the "limiter plus integrator" receiver for (c1) kappa = 30 and (c2) kappa = 1 in the presence of $S\alpha S$ noise evaluated with Monte Carlo simulation experiments.

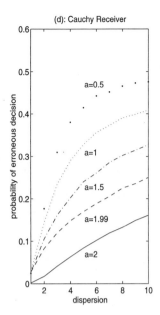

FIGURE 10.11 The performance of the Cauchy receiver in $S\alpha S$ noise evaluated with Monte Carlo simulation experiments.

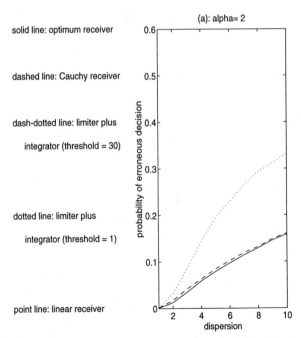

FIGURE 10.12 The performance of the optimum (solid line), Cauchy (dashed line), "limiter plus integrator" with kappa = 30 (dashed-dotted line), and kappa = 1 (dotted line), and linear (Gaussian) receiver (point line) in the presence of Gaussian noise ($a = 2$).

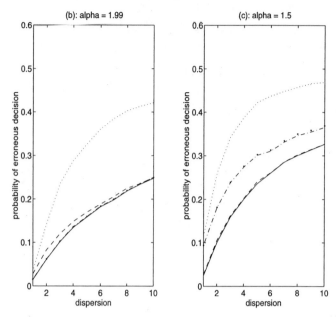

FIGURE 10.13 The performance of the optimum (solid line), Cauchy (dashed line), "limiter plus integrator" with $\kappa = 30$ (dashed-dotted line), and $\kappa = 1$ (dotted line), and linear (Gaussian) receiver (point line) in the presence of $S\alpha S$ noise when (b) $\alpha = 1.99$ and (c) $\alpha = 1.5$.

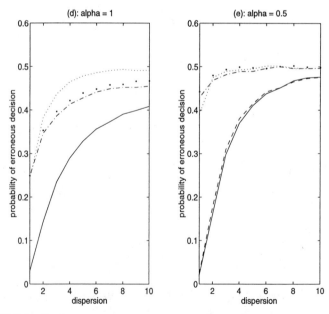

FIGURE 10.14 Performance comparison of nonlinear receivers in $S\alpha S$ noise with mismatched characteristic exponent and/or dispersion.

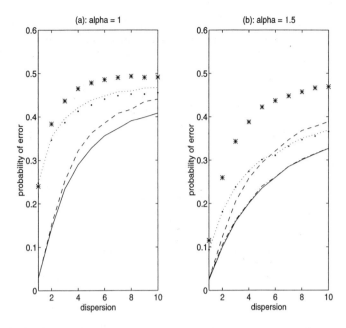

FIGURE 10.15 Performance comparisons of mismatched nonlinear receivers.

in its characteristic exponent. In Figure 10.15, we show the performance of a Cauchy receiver designed on the assumption of unit dispersion in an environment of Cauchy (Fig. 10.15(a)) and S($\alpha = 1.5$)S (Fig. 10.15(b)) noise of dispersion $\gamma = 1, 2, \ldots, 10$. The performance curves in these figures correspond to the optimum receiver (solid line), the dispersion-mismatched Cauchy receiver (dashed line), the Cauchy receiver with the proper dispersion (dash-dotted line), the linear receiver (dotted line), the limiter plus integrator receiver with $\kappa = 30$ (point line), and the limiter plus integrator receiver with $\kappa = 1$ (star line). Clearly the Cauchy receiver demonstrates robustness even in the presence of mismatches in the dispersion of the stable noise, as well as its characteristic exponent, as a study of Figure 10.15 demonstrates.

10.5 CONCLUSION

We examined the performance of four different classes of receivers in the presence of impulsive noise modeled as a SαS random process. As a measure of performance, we employed the average probability of erroneous decision as computed theoretically in the limit of a large number of samples of the incoming signal and also as obtained via extensive Monte Carlo simulations. As expected, we found that the optimum receiver outperforms the linear, the limiter plus integrator, and the Cauchy receiver in all cases in that it provides the lowest probability of error.

The traditionally employed limiter plus integrator receiver was found to perform below the Cauchy receiver or even the linear receiver in the case of low limiter threshold κ. The Cauchy receiver, on the other hand, was found to perform quite closely to, even though below, the optimum receiver for a wide range of characteristic exponents α and dispersion γ and to maintain this trend of extreme robustness even under mismatched characteristic exponents and/or dispersions.

Future research may include examining the structure of detectors that are used for the detection of impulsive stochastic transients in a background of Gaussian noise. It is also interesting to examine the possibility of constructing nonparametric receivers for robust detection in stable noise of unknown characteristic exponent and/or dispersion.

PROBLEMS

(*Note*: most of the problems are further research topics for interested readers and may be very challenging.)

1. In Middleton's strictly canonical Class A noise model, the interference is assumed to be a process having two independent components:

$$Z(t) = Z_G(t) + Z_P(t).$$

   Z_G is a stationary Gaussian background process; Z_P is a summation of waveforms from independent sources. The p.d.f. of Z consists of an infinite mixture of weighted Gaussian densities. From the APD (amplitude probability distribution), we can see that when the amplitude is under a certain threshold, the $S\alpha S$ distributions have a Gaussian behavior. It is equivalent to viewing the samples of lower amplitudes as Gaussian background noise and the samples of higher amplitudes as impulsive component. Find such threshold values for $\alpha = 1.0, 1.5, 1.8$ (assuming the dispersion is 1 for all cases). Are these values the same?

2. (Continuing Problem 1) Similarly, for the APD of the envelope of the narrowband $S\alpha S$ noise, there is also a threshold, under which the APD has a Gaussian behavior. Find such threshold values for $\alpha = 1.0, 1.5, 1.8$ (assuming unit dispersion for all cases). In the paper by S. M. Zabin and H. V. Poor, "Recursive Algorithms for Identification of Impulsive Noise Channels" (*IEEE Trans. Inform. Theory*, Vol. 36, no. 3, May 1990), the threshold-comparison method that combines the consistency of the method of moments and the efficiency of the likelihood method is used for parameter estimation. A similar idea may be applied to the estimation of the parameters of an $S\alpha S$ process.

3. Modulation classification attempts to classify the modulation type of a noise-corrupted signal. Assuming the signal is MPSK with known carrier frequency and equal power, the hypothesis testing in terms of complex envelopes for the kth signaling interval (during $kT \leq t \leq (k+1)T$) has the form:

$$H_0 : r(k) = n(k)$$
$$H_1 : r(k) = s(k) + n(k)$$

where $s(k)$ contains the complex envelope for kth transmitted symbol and the carrier phase introduced by the channel, i.e.,

$$s(k) = \sqrt{2P} e^{j\phi(k)} e^{j\theta_c(k)}, \qquad (10.36)$$

where P is the constant signal power. $\phi(k)$ is the signal phase, which takes on one of the M (for M-ary PSK) values of the set $\{\frac{2\pi m}{M}; m = 0, 1, 2, \ldots, M-1\}$. $\theta_c(k)$ is the channel phase with a uniform distribution in $[0, 2\pi]$. Now, consider L symbols are observed, show that if the channel phase is fast varying, then regardless of whether the channel is Gaussian or Cauchy, the signal phase $\phi(k)$ will be suppressed.

4. (Continuing Problem 3) Now, let us assume that the channel is slowly varying, i.e., the channel phase $\theta_c(k)$ remains constant over L observed symbols, so we can use a constant θ_c (the time index k is dropped) to represent it. What is the ML classifier of a Gaussian channel? What is the ML classifier of a Cauchy channel?

5. Derive and plot the locally optimum nonlinearity of Gaussian, Cauchy, and Pearson distributions.

6. In practice, there is always a limit on how much impulsive interference a receiver can tolerate. If we use a clipper to reject large outliers rather than increasing the tolerance of the receiver to accommodate all the collected data, how does this receiver compare with the optimal or suboptimal receivers designed according to $S\alpha S$ theory?

7. Suppose a receiver that is optimal for additive white Gaussian noise (AWGN) channel is used in an impulsive environment modeled by a non-Gaussian $S\alpha S$ law. Show the performance degradation as a function of α (the impulsiveness).

8. Examine the performance of the optimum and suboptimum receivers derived for operation in $S\alpha S$ when the noise is actually Laplace distributed.

11

Current and Future Trends in Signal Processing with Alpha-Stable Distributions

In the previous chapters of the book, we analyzed the main mathematical and statistical results that are known regarding stable random distributions and processes with particular emphasis placed on the signal processing aspects of the theory. Some of the more advanced mathematical details have been omitted, but can be found in the references or related monographs [Samorodnitsky and Taqqu, 1994]. In particular, we defined the class of stable distributions, presented their most important properties, described mathematical tools, such as the covariation between stable random variables, that are significant in the design of signal processing algorithms, considered special classes of stable random processes, developed stable statistical-physical models for impulsive noise, and, finally, demonstrated the application of these concepts to the design of signal processing algorithms via study of optimum and suboptimum receivers operating in alpha-stable, impulsive noise. The material in this book will allow practicing engineers to address other signal processing problems of the real world within the framework of alpha-stable random distributions and processes.

A number of these practical signal processing problems have, in fact, been addressed since the first draft of the book was completed. In particular, a study has been conducted of the distributions of random variables consisting of the sum of independent $S\alpha S$ random variables of different characteristic exponents [Tsihrintzis and Nikias, 1995a]. Asymptotic series expansions have been developed in Tsihrintzis and Nikias [1995a], that allow the real-time computation of the corresponding p.d.f.'s. These random variables and the corresponding processes provide generalized statistical models for mixtures of Gaussian and impulsive noise and have found application in the detection of impulsive stochastic transients over background noise.

New estimators of the parameters of impulsive interference have also been proposed by Tsihrintzis and Nikias [1994]. These estimators are based on asymptotic extreme value theory, order statistics, and fractional lower-order moments and, unlike maximum likelihood estimators, are characterized by closed-form expressions, minimum real-time requirements, and robustness to the stability assumption. The design of parameter estimators is presently being extended to the identification of FIR and IIR systems.

Blind system identification in impulsive signal environments has also been studied. When a nonminimum phase FIR channel is driven by a $S\alpha S$ random process, blind identification methods based on higher-order statistics need to be replaced by methods based on fractional lower-order statistics. When the input is a white $S\alpha S$ ($0 < \alpha < 2$) random process with unknown characteristic exponent, an algorithm has been developed to estimate the characteristic exponent α and the channel coefficients from the output data only [Ma and Nikias, 1995]. A related problem under investigation is the joint estimation of the channel impulse response coefficients and the transmitted signal in additive $S\alpha S$ noise.

A new class of robust beam formers has also been introduced. The beam formers perform optimally over a wide range of non-Gaussian additive noise environments [Tsakalides and Nikias, 1994a,b,c]. The maximum likelihood approach has been used to estimate the bearing of multiple sources from a set of snapshots when the additive interference is impulsive in nature and modeled as a $S\alpha S$ process. Transform-based approximations to the likelihood estimators were used for the general $S\alpha S$ class of distributions while the exact probability density function was used for the Cauchy case. It was shown that the Cauchy beamformer greatly outperformed the Gaussian beamformer in a wide variety of non-Gaussian noise environments, while it performed comparably to the Gaussian beamformer when the additive noise was Gaussian. The Cramér-Rao bound for the estimation error variance was derived for the Cauchy case, and the robustness of the $S\alpha S$ beamformers in a wide range of impulsive interference environments was demonstrated via Monte Carlo simulation. The optimal ML techniques employed in Tsakalides and Nikias [1994a,b,c] are often regarded as exceedingly complex due to the high computational load of the multivariate nonlinear optimization problem involved with them. Traditional subspace techniques employing second and higher-order moments cannot be applied here. Instead, properties of fractional lower-order moments (FLOMs) and covariations need to be used. This approach is presently followed and the results are expected to be announced soon.

Another avenue of research that has been followed is in the direction of derivation of robust adaptive algorithms to blindly equalize a system driven by $S\alpha S$ input. A blind equalization approach that is similar to the traditional Bussgang equalizers, such as the Godard and Sato algorithms—applicable to finite variance data—can also be derived for $S\alpha S$ data. In the filter update equation, lower-order statistics must be used because of the infinite variance nature of the data. In Lambert and Nikias [1994], such a method is proposed that uses a maximum a posteriori likelihood function to derive the correct nonlinearity needed in the update equation.

One important application of stable processes is in adaptive interference mitigation problems as they arise in signal processing fields such as communications, radar, sonar, and biomedicine. When the interference is an impulsive signal of incompletely known distribution and can be characterized as a stable process, traditional adaptive algorithms based on second- or higher-order statistics are not appropriate because of the infinite variance of the interference. A variety of new, robust adaptive algorithms for mitigating $S\alpha S$ interferences is presented in Shin and Nikias [1994] and their performance is demonstrated. The new algorithms examined in Shin and Nikias [1994] are either of the LMS-type algorithm with improved instantaneous estimates or are based on criteria using the pth norm where $p < \alpha$.

The general conditional mean (GCM) and absolute value criterion (AVC) have also been proposed for detection and estimation of impulsive interference [Shen and Nikias, 1993, 1994, 1995]. The approach is based on a general $\mathcal{L}^{(p,q)}$-metric, $p, q > 0$, on the probability space, and the corresponding optimality criterion. If this criterion is adopted for the estimation problem of complex impulsive interference in linear systems presented by state-space equations, the closed-form a posteriori density of the state (interference) can be computed recursively for both arbitrary i.i.d. state noise and any discrete-type measurement noise (multi-level complex signal), and the optimal $\mathcal{L}^{(p,q)}$-metric interference estimators can be derived for arbitrary p and q values. This new approach to the interference mitigation problem has been shown to be applicable and robust to interference of arbitrary density and its recursive nature admits both a simple theoretical analysis and a straightforward real-time implementation.

In all of the above investigations, it has been observed that the signal processing algorithms that are derived assuming a $S\alpha S$ distribution for the signals and/or the noise are generally *robust to modeling errors*, such as uncertainties in the signal/noise models. Moreover, they have been found to *outperform* the corresponding Gaussian algorithms in most cases and to maintain a performance comparable to that of the Gaussian algorithms even when the Gaussianity assumption is true. These two properties of signal processing algorithms based on $S\alpha S$ distributions, namely, robustness to modeling errors and performance higher than that of Gaussian algorithms, has really been established using theoretical arguments [Tsihrintzis and Nikias, 1995a]. As a general conclusion, we can, therefore, state that it is preferable to design real-world signal processing algorithms on the basis of non-Gaussian stable distributions and processes rather than their Gaussian counterparts to guarantee robustness when operating in uncertain and/or time-varying impulsive environments.

Appendix

In the Appendix, we summarize the basic concepts about metric spaces used in the book. For details on the subject see any standard textbooks on functional analysis.

A set X is called a *metric space* if there is a function ρ on the product set $X \times X$ that satisfies the following conditions:

(1) $\rho(x, y) \geq 0,\quad$ for all $x, y \in X$
(2) $\rho(x, y) - 0,\quad$ if and only if $x = y$
(3) $\rho(x, y) = \rho(y, x)$
(4) $\rho(x, y) \leq \rho(x, z) + \rho(z, y)$.

The function $\rho(x, y)$ is called the *distance* between the elements x and y, or the *metric* of the space X. The set X together with a metric ρ is called a *metric space*.

A sequence $\{x_n\}$ of elements of a metric space X is said to *converge* to an element $x \in X$ if $\rho(x_n, x) \to 0$ as $n \to 0$.

A sequence $\{x_n\}$ of elements of a metric space X is called *Cauchy* if given $\epsilon > 0$ there is a positive integer $N(\epsilon)$ such that $\rho(x_m, x_n) < \epsilon$ for all $m, n \geq N(\epsilon)$.

A metric space is said to be *complete* if each of its Cauchy sequence converges to an element.

A set of X is called *linear* or a *vector space* if there is an *addition* operation that maps two ordered elements x and y to another element $x + y$ in X, and a *scalar product* that takes a number λ and an $x \in X$ to an element $\alpha x \in X$ such that

(1) $(x + y) + z = x + (y + z)$
(2) $x + y = y + x$
(3) There exists a *zero* element 0 in X, such that $x + 0 = x$ for all $x \in X$.
(4) For every $x \in X$, there exists an element $-x \in X$ such that $x + (-x) = 0$.
(5) $\alpha(\beta x) = (\alpha\beta)x$ for all numbers α, β and all $x \in X$
(6) $\alpha(x + y) = \alpha x + \alpha y$ for all numbers α and all $x, y \in X$
(7) $(\alpha + \beta)x = \alpha x + \beta x$ for all numbers α, β and all $x \in X$
(8) $1 \cdot x = x$ for all all $x \in X$.

If the scalar product is defined for all complex scalars then the space is called a *complex linear space* and if it is defined for all real scalars the space is called a *real linear space*.

A *normed linear space* is a linear space X together with a function $\|X\|$, called the *norm* of X, which takes the set X into the real line and satisfies the following conditions:

(a) $\|x\| \geq 0$ and $\|x\| = 0$ if and only if $x = 0$;
(b) $\|\alpha x\| = |\alpha|\|x\|$;
(c) $\|x + y\| \leq \|x\| + \|y\|$.

Every normed linear space is a linear metric space where the distance between two elements $x, y \in X$ is defined to be $\rho(x, y) = \|x - y\|$. A complete normed linear space is called a *Banach space*.

A *Hilbert space* is a complex linear space H together with a map (x, y), called an *inner product*, which takes each pair of elements $x, y \in H$ to a complex number and satisfies the following conditions:

(1) $(x, y) = \overline{(y, x)}$
(2) $(x_1 + x_2, y) = (x_1, y) + (x_2, y)$
(3) $(\alpha x, y) = \alpha(x, y)$ for all complex numbers α
(4) $(x, x) \geq 0$ and $(x, x) = 0$ if and only if $x = 0$.

It is easy to see that the number $\|x\| = \sqrt{(x, x)}$ satisfies the conditions for norm. It is also required that H is complete with respect to this norm, i.e., H is a Banach space.

Example (Euclidean space R^n): Let $\mathbf{x}$ be an n-dimensional real vector. Define

$$\|\mathbf{x}\|_p = (|x_1|^p + \cdots + |x_n|^p)^{1/p}. \tag{A.1}$$

Then $(R^n, \|\cdot\|_p)$ is a Banach space for any $1 \leq p \leq \infty$. The norm defined by (A.1) is called the L_p norm. □

Bibliography

M. Abramowitz and I. A. Stegun, eds., 1965, *Handbook of Mathematical Functions*, New York: Dover Publications.

L. V. Ahlfors, 1979, *Complex Analysis*, 3rd ed., New York: McGraw-Hill.

V. Akgiray and C. G. Lamoureux, 1989, "Estimation of Stable-Law Parameters: A Comparative Study," *J. Bus. Econ. Stat.*, Vol. 7 (Jan.), pp. 85–93.

H. An and Z. Chen, 1982, "On Convergence of LAD Estimates in Autoregression with Infinite Variance," *J. Multivariate Analysis*, Vol. 12, pp. 335–345.

J. Berger and B. Mandelbrot, 1963, "A New Model for Error Clustering in Telephone Circuits," *IBM J. Res. and Develop.*, Vol. 7, pp. 224–236.

H. Bergstrom, 1952, "On Some Expansions of Stable Distribution Functions," *Ark. Math.*, Vol. 2, pp. 375–378.

L. A. Berry, 1981, "Understanding Middleton's Canonical Formula for Class A Noise," *IEEE Trans. Electromagn. Compat.*, Vol. 23, pp. 337–344.

J. S. Bird, 1982, "Calculating Detection Probabilities for Systems Employing Noncoherent Integration," *IEEE Trans. Aerospace Electr. Syst.*, Vol. AES-18, pp. 401–409.

R. Blattberg and T. Sargent, 1971, "Regression with Non-Gaussian Stable Disturbances: Some Sampling Results," *Econometrica*, Vol. 39 (May), pp. 501–510.

P. Bloomfield and W. L. Steiger, 1983, *Least Absolute Deviations: Theory, Applications and Algorithms*, Boston: Birkhäuser.

M. Bouvet and S. C. Schwartz, 1989, "Comparison of Adaptive and Robust Receivers for Signal Detection in Ambient Underwater Noise," *IEEE Trans. Acoust., Speech, Signal Processing*, Vol. ASSP-37 (May), pp. 621–626.

L. Breiman, 1968, *Probability*, Reading, MA: Addison-Wesley.

B. W. Brorsen and S. R. Yang, 1990, "Maximum Likelihood Estimates of Symmetric Stable Distribution Parameters," *Commun. Statist. Simula.*, Vol. 19, No. 4, pp. 1459–1464.

S. Cambanis, 1983, "Complex Symmetric Stable Variables and Processes," in *Contributions to Statistics: Essays in Honor of Norman L. Johnson* (P. Sen, ed.), pp. 63–79, New York: North-Holland.

S. Cambanis, C. D. Hardin Jr., and A. Weron, 1988, "Innovations and Wold Decompositions of Stable Sequences," *Probab. Theory Rel. Fields*, Vol. 79, pp. 1–27.

S. Cambanis and A. G. Miamee, 1989, "On Prediction of Harmonizable Stable Processes," *Sankhya A*, Vol. 51, No. 3, pp. 269–294.

S. Cambanis and G. Miller, 1981, "Linear Problems in pth Order and Stable Processes," *SIAM J. Appl. Math.*, Vol. 41 (Aug.), pp. 43–69.

S. Cambanis and A. R. Soltani, 1984, "Prediction of Stable Processes: Spectral and Moving Average Representations," *Z. Wahrsch verw. Gebiete*, Vol. 66, pp. 593–612.

J. Capon, 1961, "On the Asymptotic Efficiency of Locally Optimum Detectors," *IRE Trans. Inform. Theory*, Vol. IT-7 (April), pp. 67–71.

G. Casella and R. L. Berger, 1990, *Statistical Inference*, Pacific Grove, CA: Wadsworth & Brooks/Cole.

J. M. Chambers, C. L. Mallows, and B. W. Stuck, 1976, "A Method for Simulating Stable Random Variables," *J. Amer. Statist. Assoc.*, Vol. 71 (June), pp. 340–344.

J. F. Claerbout and F. Muir, 1973, "Robust Modeling with Erratic Data," *Geophysics*, Vol. 38, No. 5, pp. 826–844.

D. B. Cline and P. J. Brockwell, 1985, "Linear Prediction of ARMA Processes with Infinite Variance," *Stochastic Processes and Their Applications*, Vol. 19, pp. 281–296.

W. Q. Crichlow, C. J. Roubique, A. Spaulding, and W. M. Beery, 1960, "Determination of the Amplitude Probability Distribution Function of Atmospheric Radio Noise from Statistical Moments," *J. Res. NBS*, Vol. 64D, pp. 49–56.

E. Denoél and J. P. Solvay, 1985, "Linear Prediction of Speech with a Least Absolute Error Criterion," *IEEE Trans. Acoust., Speech, Signal Processing*, Vol. ASSP-33, No. 6, pp. 1397–1403.

W. H. DuMouchel, 1971, *Stable Distributions in Statistical Inference*, Ph.D. dissertation, Department of Statistics, Yale University.

W. H. DuMouchel, 1975, "Stable Distributions in Statistical Inference: 2. Information from Stably Distributed Samples," *J. Amer. Statist. Assoc.*, Vol. 70 (June), pp. 386–393.

W. H. DuMouchel, 1983, "Estimating the Stable Index α in Order to Measure Tail Thickness: A Critique," *Ann. Statist.*, Vol. 11, No. 4, pp. 1019–1031.

J. Evans and A. S. Griffiths, 1974, "Design of a Sanguine Noise Processor Based Upon World-wide Extremely Low Frequency (ELF) recordings," *IEEE Trans. Commun.*, Vol. 22, pp. 528–539.

E. F. Fama, 1965, "The Behavior of Stock-Market Prices," *J. of Business*, Vol. 38, pp. 34–105.

E. F. Fama and R. Roll, 1968, "Some Properties of Symmetric Stable Distributions," *J. Amer. Statist. Assoc.*, Vol. 63 (Sept.), pp. 817–836.

E. F. Fama and R. Roll, 1971, "Parameter Estimates for Symmetric Stable Distributions," *J. Amer. Statist. Assoc.*, Vol. 66 (June), pp. 331–338.

W. Feller, 1966, *An Introduction to Probability Theory and Its Applications*, vol. II, New York: Wiley.

T. Ferguson, 1967, *Mathematical Statistics, A Decision Theoretic Approach*, New York: Academic.

E. C. Field Jr. and M. Lewinstein, 1978, "Amplitude-Probability Distribution Model for VLF/ELF Atmospheric Noise," *IEEE Trans. Commun.*, Vol. COM-26, No. 1.

K. Furutsu and T. Ishida, 1961, "On the Theory of Amplitude Distribution of Impulsive Random Noise," *J. Applied Physics*, Vol. 32, No. 7.

A. Giordano and F. Haber, 1972, "Modeling of Atmospheric Noise," *Radio Science*, Vol. 7, pp. 1101–1123.

B. V. Gnedenko and A. N. Kolmogorov, 1968, *Limit Distributions for Sums of Independent Random Variables*, Reading, MA: Addison-Wesley.

R. Gonin and A. H. Money, 1989, *Nonlinear L_p-Norm Estimation*, New York: Marcel Dekker.

I. S. Gradshteyn and I. M. Ryzhik, 1965, *Table of Integrals, Series, and Products*, New York: Academic.

C. W. Granger and D. Orr, 1972, "Infinite Variance and Research Strategy in Time Series Analysis," *J. Amer. Statist. Soc.*, Vol. 67 (June), pp. 275–285.

S. Gross and W. L. Steiger, 1979, "Least Absolute Deviation Estimates in Autoregression with Infinite Variance," *J. Appl. Prob.*, Vol. 16, pp. 104–116.

H. Hall, 1966, "A New Model for 'Impulsive' Phenomena: Application to Atmospheric-Noise Communication Channel," Technical Report 3412-8, 66-052, Stanford Electron. Lab., Stanford Univ.

E. J. Hannan and M. Kanter, 1977, "Autoregressive Processes with Infinite Variance," *J. Appl. Prob.*, Vol. 14, pp. 411–415.

C. D. Hardin Jr., 1982, "On the Spectral Representation of Symmetric Stable Processes," *J. Multivariate Anal.*, Vol. 12, pp. 385–401.

C. W. Helstrom and J. A. Ritcey, 1984, "Evaluating Radar Detection Probabilities by Steepest Descent Integration," *IEEE Trans. Aerospace Electr. Syst.*, Vol. AES-20, pp. 624–633.

D. R. Holt and E. L. Crow, 1973, "Tables and Graphs of the Stable Probability Density Functions," *J. Research of the National Bureau of Standards—B. Mathematical Sciences*, Vol. 77B, No. 3 & 4, pp. 143–198.

J. Holtsmark, 1919, "Über die Verbreiterung von Sektrallinien," *Ann. Physik*, Vol. 58, No. 4, pp. 577–630.

Y. Hosoya, 1978, "Discrete–Time Stable Processes and Their Certain Properties," *Ann. Prob.*, Vol. 6, No. 1, pp. 94–105.

O. Ibukun, 1966, "Structural Aspects of Atmospheric Radio Noise in the Tropics," *Proc. IRE*, Vol. 54, pp. 361–367.

L. Izzo, L. Panico, and L. Paura, 1982, "Error Rates for Fading NCFSK Signals in an Additive Mixture of Impulsive and Gaussian Noise," *IEEE Trans. Commun.*, Vol. COM-30 (Nov.), pp. 2434–2438.

L. Izzo and L. Paura, 1981, "Error Probability for Fading CPSK Signals in Gaussian and Impulsive Atmospheric Noise Environments," *IEEE Trans. Aerosp. Electron. Syst.*, Vol. AES-17 (Sept.), pp. 719–722.

M. Kanter and W. L. Steiger, 1974, "Regression and Autoregression with Infinite Variance," *Adv. Appl. Prob.*, Vol. 6, pp. 768–783.

S. A. Kassam, 1988, *Signal Detection in Non-Gaussian Noise*, New York: Springer.

M. G. Kendall and A. Stuart, 1958, *The Advanced Theory of Statistics*, vol. 1, New York: Hafner.

I. A. Koutrouvelis, 1980, "Regression-Type Estimation of the Parameters of Stable Laws," *J. Amer. Statist. Assoc.*, Vol. 75 (Dec.), pp. 918–928.

I. A. Koutrouvelis, 1981, "An Iterative Procedure for the Estimation of the Parameters of Stable Laws," *Commun. Statist. Simula.*, Vol. 10, No. 1, pp. 17–28.

R. H. Lambert and C. L. Nikias, 1994, "Maximum A Posteriori Filter Estimation for Cauchy Data Using Lower Order Statistics," Technical Report USC-SIPI-253, University of Southern California, March.

R. M. Lerner, 1961, "Design of Signals," in *Lectures on Communication System Theory* (E. J. Baghdady, ed.), pp. 243–277, New York: McGraw-Hill.

P. Lévy, 1925, *Calcul des Probabilités*, Paris: Gauthier-Villars.

K. Lii and M. Rosenblatt, 1982, "Deconvolution and Estimation of Transfer Function Phase and Coefficients for Non-Gaussian Linear Processes," *The Annals of Statistics*, Vol. 10, No. 4, pp. 1195–1208.

E. Lukacs, 1960, *Characteristic Functions*, London: Griffin.

E. Lukacs, 1970, *Characteristic Functions*, 2nd ed., New York: Hafner.

X. Ma and C. L. Nikias, 1995, "On Blind Channel Identification for Impulsive Signal Environments," in *Proc. ICASSP'95* (Detroit, MI), May.

F. W. Machell, C. S. Penrod, and G. E. Ellis, 1989, "Statistical Characteristic of Ocean Acoustic Noise Processes," in *Topics in Non-Gaussian Signal Processing* (E. J. Wegman et al., ed.), pp. 29–57, New York: Springer.

B. Mandelbrot, 1963, "The Variation of Certain Speculative Prices," *J. of Business*, Vol. 36, pp. 394–419.

B. Mandelbrot and J. W. Van Ness, 1968, "Fractional Brownian Motions, Fractional Noises and Applications," *SIAM Review*, Vol. 10, pp. 422–437.

S. L. Marple Jr., 1987, *Digital Spectral Analysis*, Englewood Cliffs, NJ: Prentice-Hall.

E. Masry and S. Cambanis, 1984, "Spectral Density Estimation for Stationary Stable Processes," *Stochastic Processes and Their Applications*, Vol. 18, pp. 1–31.

W. B. McCain and C. D. McGillem, 1987, "Performance Improvement of DPLL's in Non-Gaussian Noise Using Robust Estimators," *IEEE Trans. Commun.*, Vol. COM-35, No. 11, pp. 1207–1216.

J. H. McCulloch, 1986, "Simple Consistent Estimators of Stable Distribution Parameters," *Commun. Statist. Simula.*, Vol. 15, No. 4, pp. 1109–1136.

J. M. Mendel, 1987, *Lessons in Digital Estimation Theory*, Englewood Cliffs, NJ: Prentice-Hall.

P. Mertz, 1961, "Model of Impulsive Noise for Data Transmission," *IRE Trans. Commun. Systems*, Vol. 9 (June), pp. 130–137.

D. Middleton, 1974, "First-order Probability Models of the Instantaneous Amplitude, Part I," Report OT 74-36, Office of Telecommunications.

D. Middleton, 1976, "Statistical-Physical Models of Man-made and Natural Radio Noise, Part II: First-order Probability Models of the Envelope and Phase," Report OT 76-86, Office of Telecommunications.

D. Middleton, 1977, "Statistical-Physical Models of Electromagnetic Interference," *IEEE Trans. Electromagn. Compat.*, Vol. EMC-19, No. 3, pp. 106–127.

D. Middleton, 1978, "Statistical-Physical Models of Man-made and Natural Radio Noise, Part III: First-order Probability Models of the Instantaneous Amplitude of Class B Interference," Report NTIA-CR-78-1, Office of Telecommunications.

D. Middleton, 1979, "Procedures for Determining the Parameters of the First-Order Canonical Models of Class A and Class B Electromagnetic Interference," *IEEE Trans. Electromagn. Compat.*, Vol. EMC-21, No. 3, pp. 190–208.

G. Miller, 1978, "Properties of Certain Symmetric Stable Distribution," *J. Multivariate Anal.*, Vol. 8, pp. 346–360.

J. H. Miller, 1972, "Detections for Discrete-Time Signals in Non-Gaussian Noise," *IEEE Trans. Inform. Theory*, Vol. IT-18, No. 2.

R. H. Myers, 1989, *Classical and Modern Regression with Applications*, Boston: PWS-KENT Publishing.

C. L. Nikias and A. P. Petropulu, 1993, *Higher Order Spectral Analysis: A Nonlinear Signal Processing Framework*, Englewood Cliffs, NJ: Prentice-Hall.

C. L. Nikias and M. R. Raghuveer, 1987, "Bispectrum Estimation: A Digital Signal Processing Framework," *IEEE Proc.*, Vol. 75, No. 7, pp. 869–891.

A. Papoulis, 1991, *Probability, Random Variables, and Stochastic Processes*, 3rd ed., New York: McGraw-Hill.

E. Parzen, 1962, *Stochastic Process*, San Francisco, CA: Holden-Day.

V. J. Paulauskas, 1976, "Some Remarks on Multivariate Stable Distributions," *J. Multivariate Anal.*, Vol. 6, pp. 356–368.

A. S. Paulson, E. W. Holcomb, and R. Leitch, 1975, "The Estimation of the Parameters of the Stable Laws," *Biometrika*, Vol. 62, pp. 163–170.

H. I. Pereira, 1990, "Tests for the Characteristic Exponent and the Scale Parameter of Symmetric Stable Distributions," *Commun. Statist. Simula.*, Vol. 19, No. 4, pp. 1465–1475.

S. S. Pillai and M. Harisankar, 1987, "Simulated Performance of a DS Spread-Spectrum System in Impulsive Atmospheric Noise," *IEEE Trans. Electromagn. Compat.*, Vol. EMC-29 (Feb.), pp. 80–82.

M. Pourahmad, 1984, "On Minimality and Interpolation of Harmonizable Stable Processes," *SIAM J. Appl. Math.*, Vol. 44, No. 5, pp. 1023–1030.

S. J. Press, 1968, "A Compound Events Model for Security Prices," *J. of Business*, Vol. 40, pp. 317–335.

S. J. Press, 1972, "Estimation in Univariate and Multivariate Stable Distributions," *J. Amer. Statist. Assoc.*, Vol. 67 (Dec.), pp. 842–846.

S. J. Press, 1975, "Stable Distributions: Probability, Inference, and Applications in Finance—A Survey, and a Review of Recent Results," in *Statistical Distributions in Scientific Work*, Vol. 1 (G. P. Patil, ed.), pp. 87–102, Dordrecht, Holland: Reidel.

S. S. Rappaport and L. Kurz, 1966, "An Optimal Nonlinear Detector for Digital Data Transmission Through Non-Gaussian Channels," *IEEE Trans. Commun.*, Vol. COM-14, No. 3, pp. 266–274.

J. R. Rice and J. S. White, 1964, "Norms for Smoothing and Estimation," *SIAM Review*, Vol. 6, No. 3, pp. 243–256.

G. Samorodnitsky and M. S. Taqqu, 1994, *Stable Non-Gaussian Random Processes: Stochastic Models with Infinite Variance*, New York, NY: Chapman & Hall.

L. Scharf, 1991, *Statistical Signal Processing: Detection, Estimation and Time Series Analysis*, Reading, MA: Addison-Wesley.

M. Schilder, 1970, "Some Structure Theorems for the Symmetric Stable Laws," *Ann. Math. Statist.*, Vol. 41, No. 2, pp. 412–421.

J. Schroeder and R. Yarlagadda, 1989, "Linear Predictive Spectral Estimation Via the L_1 Norm," *Signal Processing*, Vol. 17, No. 1, pp. 19–29.

J. Schroeder, R. Yarlagadda, and J. Hershey, 1991, "L_p Normed Minimization with Applications to Linear Predictive Modeling for Sinusoidal Frequency Estimation," *Signal Processing*, Vol. 24, No. 2, pp. 193–216.

M. R. Schroeder, 1991, *Fractals, Chaos, Power Laws: Minutes from an Infinite Universe*, New York: Freeman.

S. C. Schwartz and J. B. Thomas, 1984, "Detection in a Non-Gaussian Environment," in *Statistical Signal Processing* (E. J. Wegman and J. G. Smith, eds.), pp. 93–105, New York: Marcel Dekker.

J. Seo, S. Cho, and K. Feher, 1989, "Impact of Non-Gaussian Impulsive Noise on the Performance of High-Level QAM," *IEEE Trans. Electromagn. Compat.*, Vol. EMC-31 (May), pp. 177–180.

M. Shao and C. L. Nikias, 1993a, "Signal Processing with Fractional Lower Order Moments: Stable Processes and Their Applications," *IEEE Proc.*, Vol. 81 (July), pp. 986–1010.

M. Shao and C. L. Nikias, 1993b, "On Symmetric Stable Models for Impulsive Noise," Technical Report USC-SIPI-231, University of Southern California, February.

J. Shen and C. L. Nikias, 1993, "Recursive Computation of *a Posteriori* Density Functions for Arbitrary I.I.D. State Noise and Its Application to Impulsive Interference Mitigation," in *Conf. Rec. of 27th Asilomar Conf. on Signals, Systems & Computers*, Pacific Grove, CA, Nov.

J. Shen and C. L. Nikias, 1994, "General $\mathcal{L}^{(p,q)}$-Metric Estimator of Arbitrary Impulsive Interference in Linear Systems," Technical Report USC-SIPI-274, University of Southern California, Nov.

J. Shen and C. L. Nikias, 1995, "Estimation of Multi-Level Digital Signals in the Presence of Arbitrary Impulsive Interference," *IEEE Trans. Signal Processing*, Vol. 43, No. 1, pp. 196–203.

R. Shiavi, 1991, *Introduction to Applied Statistical Signal Analysis*, Homewood, IL: Aksen.

D. C. Shin and C. L. Nikias, 1994, "Adaptive Interference Mitigation Techniques," Technical Report USC-SIPI-275, University of Southern California, Nov.

M. P. Shinde and S. N. Gupta, 1974, "Signal Detection in the Presence of Atmospheric Noise in Tropics," *IEEE Trans. Commun.*, Vol. COM-22, pp. 1055–1063, Aug.

I. Singer, 1970, *Best Approximation in Normed Linear Space by Elements of Linear Subspaces*, New York: Springer.

A. D. Spaulding, 1981, "Optimum Threshold Signal Detection in Broad-Band Impulsive Noise Employing Both Time and Spatial Sampling," *IEEE Trans. Commun.*, Vol. COM-29, No. 2.

B. W. Stuck, 1978, "Minimum Error Dispersion Liner Filtering of Scalar Symmetric Stable Processes," *IEEE Trans. Automat. Contr.*, Vol. AC-23, No. 3, pp. 507–509.

B. W. Stuck and B. Kleiner, 1974, "A Statistical Analysis of Telephone Noise," *Bell Syst. Tech. J.*, Vol. 53, No. 7, pp. 1263–1320.

A. Tarantola, 1987, *Inverse Problem Theory: Methods for Data Fitting and Model Parameter Estimation*, New York: Elsevier.

H. L. Taylor, S. C. Banks, and J. F. McCoy, 1979, "Deconvolution with the l_1 Norm," *Geophysics*, Vol. 44, No. 1, pp. 39–45.

J. Teichmoeller, 1971, "A Note on the Distribution of Stock Price Changes," *J. Amer. Statist. Assoc.*, Vol. 66, pp. 282–284.

P. Tsakalides and C. L. Nikias, 1994a, "Testing of a New Direction Finding Method in Impulsive Interference Environments," Technical Report USC-SIPI-257, University of Southern California, March.

P. Tsakalides and C. L. Nikias, 1994b, "A New Robust Array Processing Method Based on the Cauchy Beamformer," in *1994 Adaptive Sensor Array Processing (ASAP) Workshop*, Lexington, MA: MIT Lincoln Laboratory, March.

P. Tsakalides and C. L. Nikias, 1994c, "Maximum Likelihood Localization of Sources in Noise Modeled as a Cauchy Process," in *Proc. MILCOM'94*, Fort Monmouth, NJ, October.

G. A. Tsihrintzis and C. L. Nikias, 1994, "Fast Estimation of the Parameters of Alpha-Stable Impulsive Interference," Technical Report USC-SIPI-270, University of Southern California, September.

G. A. Tsihrintzis and C. L. Nikias, 1995a, "On the Detection of Stochastic Impulsive Transients over Background Noise," *Signal Processing*, February.

G. A. Tsihrintzis and C. L. Nikias, 1995b, "Performance of Optimum and Suboptimum Receivers in the Presence of Impulsive Noise Modeled as an α-stable process," *IEEE Trans. Comm.*, April.

H. L. Van Trees, 1968, *Detection, Estimation, and Modulation Theory, Part I*, New York: Wiley.

A. Watt and E. Maxwell, 1957, "Characteristics of Atmospheric Noise from 1 to 100 kc," *Proc. IRE*, Vol. 45, pp. 787–794.

E. J. Wegman, S. C. Schwartz, and J. B. Thomas, eds., 1989, *Topics in Non-Gaussian Signal Processing*, New York: Springer.

E. J. Wegman and J. G. Smith, eds., 1984, *Statistical Signal Processing*, New York: Marcel Dekker.

A. Weron, 1983, "Stable Processes and Measures: A Survey," in *Probability Theory on Vector Spaces III* (S. Cambanis, ed.), Lecture Notes in Mathematics, pp. 306–364, Berlin: Springer.

A. Weron, 1985, "Harmonizable Stable Processes on Groups: Spectral, Ergodic and Interpolation Properties," *Z. Wahrsch. verw. Gebiete*, Vol. 68, pp. 473–491.

S. J. Wolfe, 1973, "On the Local Behavior of Characteristic Functions," *Ann. Prob.*, Vol. 1, No. 5, pp. 862–866.

R. Yarlagadda, J. B. Bednar, and T. Watt, 1985, "Fast Algorithms for l_p Deconvolution," *IEEE Trans. Acoust., Speech, Signal Processing*, Vol. 33, No. 1, pp. 174–182.

V. J. Yohai and R. A. Maronna, 1977, "Asymptotic Behavior of Least-Squares Estimates for Autoregressive Processes with Infinite Variances," *Ann. Statist.*, Vol. 5, No. 3, pp. 554–563.

V. M. Zolotarev, 1957, "Mellin-Stieltjes Transforms in Probability Theory," *Theory Prob. Appl.*, Vol. 2, No. 4, pp. 433–460.

V. M. Zolotarev, 1981, "Integral Transformations of Distributions and Estimates of Parameters of Multidimensional Spherically Symmetric Stable Laws," in *Contribution to Probability: A Collection of Papers Dedicated to Eugene Lukacs* (J. Gani and V. K. Rohatgi, eds.), pp. 283–305, Academic Press.

V. M. Zolotarev, 1986, *One-dimensional Stable Distributions*, Providence, RI: American Mathematical Society.

Index

L_p norm, 158
α-sub-Gaussian, 25

amplitude probability distribution, 119
autoregressive (AR) processes, 39
autoregressive moving average (ARMA) processes, 39

Banach space, 98, 158
Bessel function, 117
bispectrum, 5

Cauchy distribution, 14
Cauchy receiver, 143
Cauchy-Schwartz inequality, 48
central limit theorem, 3
characteristic exponent, 2
    definition of, 14
characteristic function
    joint $S\alpha S$, 31
clipper nonlinearity, 134
complex $S\alpha S$ random variables, 53
converging variance test, 63
covariation, 46
covariation coefficient, 47
covariation matrix, 80
cumulants, 5

detection probability, 131
dispersion, 7
    defined, 14

Euler's constant, 69

factorial moment-generating function, 112
false-alarm probability, 131
fractile, 58
fractional lower-order moment, 23, 32
fractional lower-order moment estimator, 72

gamma function, 16
Gaussian distribution, 14
Gaussian mixture model, 109
generalized central limit theorem, 3, 22

hard-limiter nonlinearity, 134
heavy tail, 17
Hilbert space, 37, 98, 158
hole puncher nonlinearity, 134

impulsive noise, 109
in-phase component, 127
inner product, 158
isotropic α-stable, 116

least absolute deviation (LAD), 82
least-squares criterion, 4
linear regression, 50
linear regression property, 53
linear space, 157
    complex, 158
    normed, 158
    real, 158
LMAD algorithm, 103
LMP algorithm, 103
LMS algorithm, 103

local optimality, 132
locally most powerful detector, *see* locally optimum detector
locally optimum detector, 132
log-tail test, 63

matched filter, 132
metric space, 157
   complete, 157
   metric of, 157
Middleton's Class A model, 123
Middleton's Class B model, 123
minimum dispersion criterion, 7, 99
ML receiver, 145
moment theory, 3–7
   fractional lower-order, 6–7
   higher-order, 4–6
   second-order, 4
moving average (MA) processes, 39

n-fold dependence, 25
negative-order moment, 34
negative-order moments, 66
normal probability plot, 63

order statistics, 58
orthogonality principle, 100

parameter estimation
   maximum likelihood, 57
   sample characteristic function method, 61
   sample fractile method, 58
Pearson distribution, 14
power function, 131
power spectrum, 4
pseudo-linearity, 48

quadrature component, 127

sample characteristic function, 61
screened ratio estimator, 74
second-order moments, 4
singular value decomposition, 90
skewness function, 24
spectral measure, 24
stability property, 20
stable distribution
   applications, 7–11
   characteristic exponent, 14
   characteristic function of, 13
   density function of, 14
   location parameter, 14
   multivariate, 24
   scale parameter, *see* dispersion
   special cases of, 14
   standard, 14
   symmetric, 14
   symmetric parameter, 14
stable law, *see* stable distribution
stable processes, 37–43
   harmonizable, 42
   linear, 38
   sub-Gaussian, 37
sub-Gaussian, 53
symmetric stable distribution, 14–19

trispectrum, 5
truncated mean, 60

uniformly most powerful detector, 132

Yule-Walker equation
   generalized, 80
   higher-order, 88

zero-memory nonlinearity, 133